Proud to Be: Writing by American Warriors
Volume 10

Southeast Missouri State University Press • 2021

Proud to Be: Writing by American Warriors

Volume 10

Edited by James Brubaker

Partners in the Military-Service Literature Series

MISSOURI HUMANITIES

SOUTHEAST MISSOURI STATE UNIVERSITY · 1873

Proud to Be: Writing by American Warriors, Volume 10

ISBN: 978-1-7330153-5-6

First Published in the United States of America, 2021
Southeast Missouri State University Press
One University Plaza, MS 2650
Cape Girardeau, MO 63701
http://www.semopress.com

Cover photograph: "Patrol" by Christian Tillett
Cover design: Carrie Walker

Southeast Missouri State University Press, founded in 2001, serves as a first-rate publisher in the region and produces books and *Big Muddy: Journal of the Mississippi River Valley.*

The Missouri Humanities Council is a 501(c)3 non-profit organization that was created in 1971 under authorizing legislation from the U.S. Congress to serve as one of the 56 state and territorial humanities councils that are affiliated with the National Endowment for the Humanities.

Contents

Bill Glose

Foreword: Owning Your History

I was once a soldier in the 82nd Airborne Division and my closest friend and writing mentor, Billy Walsh, was a Marine. Naturally, we've spent many hours trading barbs, our mocking digs frequently laced with vulgarities and focusing the ineptitude or miscreant behavior typical of the other's branch of service. Harsh as our words were, they were grounded in mutual respect. Whenever the subject of military service comes up, Billy always says the same thing: "I count no man less for having never served, but those who have (pause) I count twice."

The nature of an all-volunteer military is that everyone in uniform swears a solemn vow to serve their country. To many Americans, *serving your country* is a generic patriotic term that could be replaced by *playing baseball* or *eating apple pie*; but to those who actually *served*, who stepped into a recruiter's office and signed their names, who pledged willingness to fight for their country and die if necessary, *serving your country* represents an ethos of putting others before yourself. And once you join a unit, you find the same strong character in comrades, whether they hail from a big city or a rural speck, whether their parents were well-to-do or dirt poor, and whether they come from the same ethnic background as you or one that is wildly different. No matter what past history brought us to the military, we were all the same after we arrived. As my basic training drill sergeant used to say: "We all bleed green."

After the initial choice, after signing your name on the dotted line, all other choices are gone. From that point on, you do what you're told. Another thing my drill sergeant loved to do was ask for volunteers; before anyone in formation could reply with a Hoo-ah (not that anyone would), he would point to a couple of soldiers and say, "You and you, step forward. What, did you forget you already volunteered?"

Basic training does a great job of taking the soft clay of our civilian selves and molding it into something hard, something usable. We become cogs in the military machine, each with a specified task that serves the overarching purpose: to be prepared to fight and win battles. Being combat-ready meant that we'd been trained how to function as a member of a cohesive unit in the chaos of war but not how to function as an individual afterwards, how to make sense of everything we'd seen and done, how to keep on keeping on and live happy, productive lives. Afterwards—*after war*—you have to balance the shiny ideals you once embraced with the muddy truths you actually experienced. And that can be just as scary as going off to war in the first place.

When my unit came home from Iraq, each soldier reintegrated into society in his own way. Combat pay had nested and multiplied in bank accounts while we were in the desert, so many returning vets bought new cars and rounds for everyone at the bar. Fights were frequent. All that stoked up fuel in our veins wanting to burn everything down.

I took after my father, a Vietnam vet who never spoke about his war. Stoicism was his fortress. It became mine too. Silence served as a tolerable stopgap when action ruled my world, but once I left the Army, I had too much free time to *think*. Reports about the scores of veteran suicides had me worried. I hadn't considered taking my own life, but some dark vortex had me in its grip. My choices were simple: I could follow the lessons of my silent father or try something different. And so I wrote about my experiences.

At first, I focused on tiny moments of Army life that bore little consequence—soldiers breaking down weapons and cleaning latrines, stuffing pictures of girlfriends in the webbing of Kevlar helmets, trading MREs with Bedouins for rides on their camels. I wrote about the quiet camaraderie of paratroopers huddled together with rifles on their laps. I wrote about the desert's rock-strewn plains, as foreign to us as the dark side of the moon. I wrote about a shrieking blare one night that sent us scrambling for our protective masks, only for the Nuclear-Biological-Chemical team to discover a goat had peed on the air-sampler alarm.

From those mundane moments I inched toward memories freighted with remorse. I wrote about the husks of bodies in back-splintered bunkers and wounded soldiers crying in medevac helicopters. I wrote about ripping patches from the sleeves of dead men and carrying them home as trophies. I wrote about the children we searched at gunpoint, the fear in their eyes, the worry that we might execute them. I rummaged through those things I'd bottled-up inside—my own fears, guilt, and shame—and found the exploration cathartic. Each new poem or essay felt like a scab being picked away. Each time I ripped it off, the scab shrank, the pain diminished, and the skin beneath grew healthier.

For years, I shoved my scribbled work into drawers. Sometimes I'd revisit those pages and cut into my words with the scalpel of my red pen. I didn't keep them hidden away because I was seeking perfection; I was simply too scared to let them out in the world. Writing about my experiences had been hard enough; exposing them to the public seemed impossible. If people glimpsed my private thoughts and knew what violence lurked inside my seemingly civil body, surely they would think me a monster.

I started attending open mics at bookstores and coffee shops. At first I came to listen, to see what other people were doing with their words and to figure out how mine stacked up. The lineup was a hodgepodge of homilies and rants, songs and silly cat poems. But whenever someone shared a painful parcel of their personal geography—rapes, addictions, disease, dementia—the crowd's support was overwhelming. The universal response was one of empathy.

After several months, I worked up the courage to share one of my war poems. My voice quavered as I read my work, the page trembling in my hands. A skewer of raw emotion stabbed through me and I nearly broke down crying, shocking myself as much as anyone else. I'd roared and exploded in front of others before, but never wept.

I'd always imagined people reacting to my poems with the wrinkled nose and snarled lip of disgust or else the raised-and-tightened lip-corner of contempt. But my dread had been unfounded. No one recoiled in horror. When I finished, the audience thundered their applause. Later that night, a few people offered thanks and shared stories of sons and daughters in similar situations. The relief I felt in knowing I wasn't alone, that I wasn't viewed as some alien abomination, was immeasurable.

Trauma invades everyone's lives, soldiers and civilians alike. These people, I learned, had their own troubled backstories, their own apprehensions. Each was also seeking acceptance from the group. And one by one, as we spoke into the mic, we faced up to our damaged histories and liberated ourselves from their dominion. If you don't *own* your history, your history will own you.

Far too long I'd fretted over my past, allowing it to rule my present. But after receiving such acceptance, I began submitting poems, stories, and essays to journals and found acceptance on a wider scale. In 2013, FutureCycle Press published a collection of my war poems titled *Half a Man*. For the cover picture, they used a picture of me in Iraq, young, grim-faced, and holding an AK-47. I wish I could go back in time and tell that kid that nothing troublesome ever gets better by ignoring it. "Do your job," I'd say. "And when your job is over, open your mouth and speak. Talking is not a sin."

What I've discovered in writing about my war is that words have the power to heal, and that healing can be shared. It's like a form of group therapy. And that's what *Proud to Be* provides, a community in which to gather and share our experiences.

Being part of a veteran's writing community gives me a sense of kinship I haven't felt since being a soldier. I feel like I'm still fighting for the soldier, sailor, airman, or Marine in the figurative foxhole next to me. Speaking my truth and listening to theirs, our words have the power to make us whole.

I've been a contributor to *Proud to Be* since its first year, and I've been submitting to it ever since and reading with great interest the multifaceted stories, poems, essays, and interviews by veterans and their families. The tale of an Army chaplain being screamed at in the streets of Salt Lake City, of a daughter getting to better know her Marine father while teaching him yoga, of a sailor competing in an impromptu swimming competition in Bahrain. In many ways their experiences were nothing like mine and, in a universal way, exactly the same. These glimpses of my comrades' lives let me know that whatever happened in the past, the future can still be anything I want to make it.

Fiction

Fiction Winner

Benjamin Inks

On a Mountain in Logar

I.

Nothing was the same after we took incoming on OP Tiger Eye.

It was just before lunchtime. I was up on that mountain burning shit in a barrel when the mortars began to land. Before impact, I thought I'd heard the hollow "pop" of distant artillery, but hunched over, stirring a molten stew of human fecal matter, I placed little value in my senses. My eyes burned; I didn't dare open my mouth. Breaths were taken in short, controlled bursts, craning my neck sideways like a long-distance swimmer.

Sergeant Holloway stood about twenty meters back, supervising. Even at that distance he wore disposable blue gloves, just in case he might have to touch something associated with this task.

"It needs more gas—more gas!" he called to me as I worked the charred two-by-four clockwise. "Oh for the love of—" He threw his hands in the air and began striding toward me with the look of a disappointed father about to ignite a charcoal grill for a bumbling son.

This was when the first one struck and our deployment changed forever, but perhaps I'm getting ahead of myself . . .

Our cyclical two weeks up on OP Tiger Eye always felt like a mini vacation from the massive corporatey FOB we were posted at. If FOB life was an extended glamping trip—with showers, internet, and even fast food—then OP Tiger Eye was like roughin' it backcountry. We bathed ourselves with baby wipes, we ate strictly MREs, and our entertainment was our comradery, which often took the form of a lamp-fire ghost or girl story, fabricated or fact. We couldn't build a real fire at night for fear of outing our position, so a dim red headlamp was placed between us in lieu of real warmth. This too was eventually deemed a risk, and then we told our stories in the dark.

By far the best storyteller was Specialist Grayson who had some college under his belt in Film Studies. We enjoyed listening to him, but we couldn't trust his authenticity. It always felt like he was testing out Film Studies craft on us, constantly deviating from the truth for the sake of spinning a more compelling yarn. The best lies are always an elaboration on simple truths.

Staff Sergeant Holloway was the more sincere storyteller, not to be confused with *the better*. This guy also understood the art of story. He had joined the Army before 9/11, back when the military was preparing for Russian hordes, not insurgents hidden amongst shepherds. These two bards would often go tit-for-tat like a rap battle—first something youthful, vicarious, and larger than life, and then something earnest and pertinent to our deployment.

One story that left a permanent bruise on my super-soldier ego was Sergeant Holloway's "Grenade" story. This story terrified me because of its honesty. It made me confront emotions I was doing well at the time to avoid.

By mnemonic proximity, I also remember the tale Grayson told that night; it wasn't his best, but much like Holloway's "Grenade" story, it might have been his most honest. Grayson's story also impacted me, impacted everyone, although its effects wouldn't be felt until much later in the deployment.

Newly twenty-one, Grayson was rolling down I-5 at 0200 after partying at a cowgirl-themed club in downtown Sacramento. He was annoyed because he was stone-cold sober, transporting a Honda Civic-load of drunken college kids. One plucky tagalong in the back, affectionately referred to as Rumble-belly because of his husky physique, with no warning, with no request to pullover, rolls down the window and spews a stream of vomit. At sixty miles per hour, it was like unfurling a liquid banner of pink ribbons. The side of Grayson's car was painted by it. Rumble-belly, the poor sap, had been duped by the scantily clad cowgirls into consuming a copious amount of strawberry-flavored tequila. Ten bucks a shot, but they all "wooed!" while he tossed them back. To his credit, this poor kid, with his empty wallet and his empty stomach, was more than apologetic. "I promish I'm gonna clean it for you, man. I'mma sooo sorrry." This placated Grayson, and they all shared a good laugh at Rumble-belly's expense. Then the blue lights popped in his rearview mirror. They all thought they were instantly fucked, in the way everyone panics and assumes guilt when confronted by authority. But taking inventory, they realized they had done nothing wrong: everyone was twenty-one; Grayson hadn't touched a drop; they had no booze in the car; he was going the speed limit. That's when he noticed the trooper's wipers ticking left-right in double-time, fanning streaks of pink. The trooper's head was out the window as he came to a stop, that's how smothered his windshield was. He stormed up in that trooper-drill sergeant hat and probably did some shit that was illegal. Grayson and Rumble-belly were manhandled, and everyone was made to stand against the guard rail. The car was tossed: glove box, center console, trunk. Try as he might, there was nothing to arrest them on. Grayson blew a 0.0%; no one had any warrants. The more the trooper investigated, the more volatile he became. Finally, Rumble-belly, in classic obsequious form, broke down and offered to clean the trooper's windshield. But with what? His shirt! His favorite Abercrombie shirt he bought because he thought it might improve his odds with the ladies, striped in blue and white. He stripped it off and his pointy, corpulent man-breasts swayed as he mopped up his own vomit. The trooper didn't exactly tell him to do this, but he didn't stop him either. When it was done, and they were released, Rumble-belly sat quietly in the back, bare-chested with his precious shirt stinky and balled up on his lap.

"So let that be a lesson to you, gents," Grayson said after finishing the story. "Always do the right thing. Even when you're doing the wrong thing."

As a combat arms unit, we had our fair share of DUI's, of bar fights, of run-ins with the law. Extrapolation on this phenomenon could be its own social study. Grayson's story almost had a cautionary edge to it. Something that might be told during a safety briefing. *Okay, you numbskulls—don't beat your wives, take it easy on the dog, but above all—do not—I repeat—do not—drink and drive. . . . Now enjoy your three-day weekend.*

But then Grayson said something that changed not only the dynamic of his story but how we viewed him.

"But I must confess I haven't been entirely truthful," he said to us all. "For I told you I was the cool, confident driver bearing witness to quiet little Rumble-belly's humiliation. In fact—it was the reverse. Yes, my friends, fat little Rumble-belly *was me.*"

Holy fuck. This was inconceivable. *Grayson used to be fat and awkward?* Why was this so tantalizing? I couldn't picture it. He wasn't a swole PT stud like Sergeant Holloway, but he was more than competent hiking around the steep terrain. He had an easy smile, surfer dude vibes, and was a go-to lower enlisted leader, despite not having much time in service.

"You're lying," someone just as perplexed as I called him out.

"No, it's the God's honest truth!" said Grayson, right hand raised.

We all thought the same thing: this reversal was a narrative twist, maybe something he snuck in last minute. Stories like this always hit harder in the first person. Grayson was just testing out new material. A bait-and-switch. We thought we were listening to a story on morals, but really it was a story of an underdog, which Grayson clearly was not.

Plus, we had all heard this story a thousand times before. The track and field coach who had knock-knee. Rocky Balboa. Arnold Schwarzenegger. Grayson's story was too archetypical. It would have been believable if injected in casual conversation: "My waistline used to be forty-two, but look at me now in these Army greens!"

Passed off as a narrative, no one believed the college-educated and highly capable Grayson was once someone shy and awkward like Rumble-belly, but we enjoyed his creative storytelling all the same.

II.

I should probably articulate the confusing origins of OP Tiger Eye. Much like Grayson's story, this too is going to sound like exaggerated reality, a picture-perfect fairytale arc: too coincidental to be true. But I assure you, to the best of my recollection—this is ninety percent real.

During pre-deployment planning, our unit believed OP Tiger Eye was named for its perfect oval shape on a topographic map, with the dozens of squiggly lines radiating from its peak resembling tiger stripes but also clueing us in as to how much a pain in the butt it would be to summit it, literally and figuratively. Let's just say you needed the eye of the tiger to conquer this beast. But would you believe an actual feline resided with us on the mountain? If that surprises you, imagine our surprise when we first occupied the area and discovered a cuddly orange tabby living with the soldiers.

We had just hiked up for the first time—and already I wanted it to be the last time.

"Gentleman, welcome to OP Tiger Eye," a staff sergeant from our counterpart unit said. He had a cheerful, peppy voice. He was going home; we were just arriving.

We pretty much collapsed at his feet. With our rifles, our rucksacks, our Kevlars and plate carriers, hiking up to OP Tiger Eye was no joke. We had trained for this, but it still broke us off like a stale KitKat bar. Anytime you see some high-speed, twenty-year-old ROTC kid on the Stairmaster in the gym wearing combat boots and plate carriers, he's training for the same thing. All I can say is Godspeed there, fella.

So we were taking a tour of this mountain, drenched in sweat, our bodies so starved of sodium that warm water from our Camelbacks tasted like Mountain Dew, and this happy red cat emerges from a supply shack and rubs against the staff sergeant's calves. We all thought we were hallucinating. It was unreal. An obvious glitch. Something from one world spliced into another. Like seeing a Goombah from *Super Mario Bros* waddle across a *Sonic the Hedgehog* game. This just shouldn't be. But it did. It did be.

"No one knows how he ended up on the OP," this sergeant explained to us. A couple of his joes had it in mind to name him Jack Bauer. Not after the show *24* but apparently after a legendary episode of *It's Always Sunny in Philadelphia* where the gang adopts a junkyard tomcat and gives him this name. His soldiers ran around for about a day calling the cat *Agent Jack Bauer* before they realized that the words *Home of Chester* were spray-painted in Redrum font on the back of the supply shack. After a few rounds of back-and-forth deliberation, taking into consideration that he responded to *Chester*, they went with this name. And so did we.

He was beautiful, with murky green eyes, flaring ear tufts, and distinctive stripes intimating a bit of wild left in him from his distant ancestors. Very fitting, for this petite and playful kitty, who in another life could have lived a plush existence as lap cat to my mom, was a ruthless killer. Every other day we would find a new bird carcass deposited somewhere on the OP. But despite his murderous proclivity, he was a loving little critter who warmed our souls by the very brightness of his fur.

Living outdoors in such primitive conditions, our minds naturally grew more spiritual. I'd often find myself stargazing at night, imaging a genesis for OP Tiger Eye. Before we came with our airlifted Connex container acting as a barracks, before we marched up with our optics and sandbags and belts of grenades for the Mk 19, before all this there was only a mountain and a cat. We were here now—as guests—but before us there was only Chester. It was clear that this mountain belonged to him.

III.

Doing my absolute best to remain true to his likeness, Sergeant Holloway's "Grenade" story went something like this:

I did Iraq in '03. Now that was a war. Shock and awe, baby. None of this pussyfootin' around counterinsurgency BS. It was glorious to see the full might of the U.S. Military unleashed. Glorious but also terrifying to live under the constant threat of mayhem. Everything felt precise and coordinated, yet also unexpected. "Christ!" I remember thinking after seeing a 1,000-pound JDAM go off: "I didn't think it would be that bright." You always have an expectation of how something's going to be, and then you have reality, but there's also this chaotic middle ground. That's where I existed when they first handed out grenades.

Our supply guy brought them out in bulk with a clipboard, just like any other piece of issued equipment. We even had to sign for them back then. I was like: "So if I pop this sucker and have nothing to clear supply with—what then? Will I owe the Army money?" My squad leader told me not to overthink it.

"Just have it on your person and pray you never have to use it," he said.

They gave us all the same little pouches we have now to secure it to our flak jackets. Only the flak jackets of yesteryear aren't nearly as high-speed as the IOTV's today. I couldn't place that grenade nowhere without it being a bother. Down low worked well until you dropped in the prone to shoot, then you had this apple-sized bomb digging into your appendix. I eventually settled on having it up high near my heart, where it was in constant view of my periphery. It was like having a second nose, it was always in sight, unavoidable no matter where my head turned. But unlike my nose, I couldn't seem to ignore it long enough to forget that I had it. It became this fraught alien appendage, something I grew increasingly worried about as the war progressed.

Now, a grenade, like a bayonet, is almost an antiquated item. You can't throw a grenade very far. At best, you'd chuck it in a bunker—or more likely someone's living room—and hope it does the hard part for you. This was probably what my squad leader meant by saying "pray you never have to use it." If you whip out a grenade in a firefight, that is some desperate close-quarter shit you got yourself into. None of this was lost on me as I carted my grenade around from block to block, city to city, its sheer weighty hang reminding me that at any given second—something had the potential to blow up in my face.

Maybe I'd have been more confident had I received the proper training. But to this day it's a dirty secret that I never qualified with grenades. Yes, you heard that right. I swam through the course, just like all y'all, no doubt. Then it came to the live-throw, and I was ejected from the pit after my first toss. See, one of the drill sergeants had placed his green, big-ass hat on the middle mannequin, and all the troops wanted to stick their grenade closest to that target. When I threw mine, instead of ducking behind the protective wall, I watched open-mouthed as my grenade landed in the patch of shade made by the drill sergeant's hat. Maybe I wanted to watch that stupid hat blow up, maybe I was stunned at such a lucky toss. I must have stared a little too long, 'cause I found myself slammed against the wall by the supervising cadre who pressed over me like he was saving my life. I didn't get to throw my other two grenades to qualify. I thought I was a no-go, a failure. Well, they must have needed fucking soldiers—and stat—'cause I was pushed through to graduation without a word about it. They couldn't afford to reject or recycle anybody after 9/11.

Here's the funny, true part: the drill sergeant's hat was no worse for the wear. Literally. He dusted it off and stuck it back on his fat, bald head right after I'd blown it skyward and six feet back. It made me think: "What good are these things? Other than making such an awful racket."

So, is there such a big difference between throwing one grenade and three? Well, it took Goldilocks three tries to find the right porridge, the right bed. I was strapped with something on my chest too hot and too hard for me to be comfortable with. My heart literally beating against it. We were seeing some real shit, too. On some occasions, enemy bullets were close enough to communicate their "crack" or their "bizz" or their "thump" or their "sizzle." We were quickly becoming fluent in the many dialects of death, yet I couldn't shake the anxiety of my own grenade. I just knew the day would come when I'd get shot in the grenade, or I'd roll over on it wrong, or intense desert sunlight would superheat it and blow my head off, cartoonishly intact and pouting like the remains of a suicide bomber.

Then one day on patrol, we take contact. By the time we find cover we realize it's coming from above. So we start lighting up this rooftop-alcove thingy, but we're at the wrong angle, chipping at a wall while the insurgent points his AK down and fires blind. My squad leader shouts for me to arc a grenade while everyone else suppresses fire.

Let me just pause and say to all y'all privates: this is a bad idea. Usually you throw "from" the rooftop, that way you're not fighting gravity. But whatever, pinned down, being shot at, being yelled at, I take my stupid grenade—pull pin—and hurl it. It doesn't make it. It clips the lip of the rooftop, bounces back and lands right between my feet. I freeze, dumbfounded. I almost feel like laughing. I knew this fucking grenade would kill me. That's when I get tackled from behind. Basic training all over again, only this time it was one of the most selfless acts I've ever witnessed. My squad leader smothers me. We stay like this cringing and flexing and gnashing our teeth for five seconds . . . ten seconds . . . and nothing. The grenade was a dud. After we call in an airstrike for the shooter on the rooftop, we learn to never speak of this moment again.

My squad leader's risky decision to have me throw a grenade "up" was negated by the sacrifice everyone saw him make. People considered putting him up for an award, but then we'd have to document the whole debacle in greater detail. Back at Saddam's palace, I kick my feet up and realize that this whole time I was afraid of nothing more than a paperweight. Admittedly, it felt good to be rid of it. To have this empty camo pouch that once housed such paranoia. That's when the supply guy comes around—having heard about our exploits, he issues me a shiny new grenade. And yes, I have to sign for it. This time, though, I'm not quite so worried. I'd thrown two. If the first one was too hard and hot, and the second one too soft and cold, I figured if I had to pop this new grenade, it would have turned out ju-u-u-st right.

But he never threw a third grenade. Not because he was afraid, but the occasion never arose. "I'm the sergeant, but all y'all privates have still thrown more grenades than me," he openly admitted. Even though that was true, he *had* thrown a grenade in combat. We all looked at our own grenades a little differently after his story. Especially me as the squad's grenadier. I carried not only a hand grenade but a bandoleer of 40mm rounds around my waist like a cop belt. A walking, talking, gum-chewing IED. But not an Improvised Explosive Device, an *intended* explosive device. Sergeant Holloway's stark honesty differed from the usual tune the Army was whistling at that time. Advice we were either told explicitly or arrived at ourselves.

Even if I wanted to be concerned about the ordinance I wore, I know what most leadership would have said.

"Don't think about it, Private. Pretend you're invincible. You're Captain Fucking America, aren't you? You can't die. Not from your own grenade. What kind of a piss-poor story would that make?"

IV.

Whenever we cycled away from OP Tiger Eye and ran regular missions out of the FOB, I always felt an inner longing for that mountain and that cat. Features I was becoming increasingly possessive of: *my mountain, my cat.* I just couldn't wait for the next few weeks to end, until it would be our turn up there again. I thought about what it would take to rehome Chester to the FOB, where we would be able to care for him better and he would have the opportunity to enliven so many other troops. This was impossible, however; his existence was already contested.

When First Sergeant found out a cat was up there with us, he commanded over the radio for us to "Get rid of it." The warming weather would make Chester a flea magnet. Plus, we didn't know what sorts of diseases he might already be carrying.

Sergeant Holloway looked at me, radio mic in hand, and put his career on the line for my cat. "I'm not sure that would be wise, Top. He's doing a good job keeping the mice out of the MRE's."

"The mice? On a mountain?"

"Roger, Sergeant. The mountain mice. Yeah, they're up here."

Nothing to say of the effect Chester had on our morale, which wouldn't have contributed much to the conversation.

Brief static over the radio, before First Sergeant said, "Fine. Whatever. Just don't none of you get bit by him. If that happens, I'm hiking up there to kill him myself."

With First Sergeant's threat out in the open, I was never confident Chester would be there when we returned. I would sit at the FOB and worry some dumb private from second platoon would antagonize him—get bit—and that would be the end of it. Fortunately, I think my brothers relished this spunky kitty about as much as I did. In the end it would take something much worse to separate Chester from his mountain.

My mom was delighted to learn we had a cat in Afghanistan. So amused by this fact that the next care package she sent included more creature comforts for Chester than for me. This was how I earned my inevitable call sign. See, we would always open our mail Christmas style, really draw it out to make the most of it. This way we could also consider what trades to make. "One of my Copenhagen's for one of your Marlboro reds? . . . Yo, can I get that copy of *Guns & Ammo* after you? . . . No way—Grayson got Gushers!"

On this particular occasion, I cut into my long-awaited care package greedily expecting a new game for my Nintendo DS and a month-long supply of jalapeno beef jerky. Instead I found the latest issue of *Cat Fancy* magazine and a dozen cans of Fancy Feast. The same look of hysteria spread across everyone's face after flipping the cardboard flaps to reveal these commodities swimming in a bed of pink packing peanuts. By the time the laughter ceased, there wasn't a dry eye in the hooch.

At lights-out, when the whispers of conversation started snuffing out, Grayson called out in the dark, "Good night, Fancy Cat," and ignited another round of laughter. I laughed too, believing it to be a one-off. One final nightcap to an evening of jokes at my expense. When Sergeant Holloway greeted me the next morning with "Good to see ya, Fancy Cat!" I knew this was now my designator and there was nothing I could do about it. You don't choose your own call sign, anyone who tries is lame. *Can I be Arrow?* I remember one new guy asked, after Green Arrow, the comic book character. We called him *Green Peen*, instead.

So my new name with all its derivatives was okay by me. I went to sleep that night an unknown PFC with the boring role-based nomenclature of 1-7D, but I awoke reborn as *The FC.* The Foxtrot Charlie. The dude who sacrifices his own junk food to feed a stray animal. The Fancy Cat.

V.

Twelve cans of wet food equated to thirty-six extra ounces packed inside my rucksack, which felt like thirty-six *pounds* halfway up the mountain. Maybe it was delirium from exhaustion—my brain flooded with endorphins—but I was sure I'd attained some measure of enlightenment after suffering to lug all this up to Chester. He devoured the first tiny can of beef and poultry with gusto. Licked his chops after, unashamedly approving of American consumerism.

I was a needle-thin cog in a vast machination of war and order. But if I could do this one small thing—feed and care for this animal—then within my humble means I would have fostered goodness. But what if I could do more than that? What if I could get him back to the States somehow? You hear stories all the time of soldiers on deployment rescuing animals. Admittedly, this was selfish romanticism, as there had to be a local shelter or nonprofit in the country that would have taken him. Probably all the way in Kabul, but Kabul is a hell of a lot closer than Kansas. Although Kabul strangely seemed like the more unfeasible of the two. Imagine running that mission by the first sergeant? *Yes, I would like to use manpower, resources, instruments of war to escort a tabby a few hours north to the capitol of this rugged nation. Maybe we can make a PR thing out of it. Think of the book and movie options. . .*

But I never had the opportunity to rescue Chester. On the day in question, the first mortar landed nowhere near Holloway and me, but its impact was still scarier than anything I'd experienced thus far. We both ate dirt fast and stared at each other in the prone, the fear in our eyes burning like sparklers in cold December. A partly vicarious fear. See, the latrine was a decent hike outside of the OP, for obvious reasons, but the round struck dead-center of the compound. It would have been preferable for them to hit me and Holloway; then they'd at least have to walk their rounds to the real target. Somehow, these fuckers scored a direct hit on their first guess. A direct hit on eight of my brothers. My sympathetic nervous system went on overdrive, stimulating nerve bundles around my solar plexus to shunt blood from my stomach to my legs that I might fling my weapon aside and sprint for the relative safety of the opposite direction. But this sensation was overpowered by an empathetic fear for the safety of others. A very different sort of dread housed in my throat and urging me not away from—but toward the danger.

We'd all experienced indirect fire at the FOB before, but that was almost passive compared to being on this dime-sized tip of mountain. If a round even found its way onto the FOB, there were bunkers, alarm systems, and vast open spaces where most ordinance would explode with menace but without harm. Teams stood by to provide counterfire, and the only responsibility for people like myself was to act like a seasoned Midwesterner staring down a tornado-green sky on their porch with a can of PBR—wait and inevitably the danger would pass.

On the OP, however, we were alone and limited in our response capabilities. Fortunately, Sergeant Holloway's first command after kicking into gear was to send me to higher ground, up the only jagged slice of mountain taller than the OP. "Get the fuck up there and see what you can see," he said. This felt like an acceptable middle ground to the conflicting flight-*no*-stay-and-fight phenomenon my body was grappling with. Holloway, the guy who openly admitted a phobia of spontaneous explosions, bravely sprinted toward the OP. Now, you have to believe me when I say—I don't have the best recollection of what happened next. I wasn't in the midst of it, of any of it, really. But apparently, the boys at the OP didn't require my eagle eye to identify the origin. Holloway and a quiet, pasty-skinned PFC who wore forever-tinted lenses and who no one expected very much of, together they slung hundreds of 40mm rounds Rambo-style from the Mk 19 until our resident mortar man could respond with artillery of his own.

The memories that are distinct are the images afterwards. Descending to the OP, smoke and little baby fires actively smoldered like the slopes of Mount Doom. And hearing Grayson cry out in agony.

"Aw fuck! I'm fucked—Jesus!"

He was on his ass, picking at his plate carrier, and then tugging at his belt, trying to remove either, not realizing he can't do both at once. A medic rushed to him, sling bag spilling out supplies. I joined too, then another. We three pulled the safety lanyard and his IOTV fell apart like it was only every male stripper wardrobe at a bachelorette party. We always thought this was a funny feature. Sometimes we yanked these lanyards as a prank, like flicking a dude in the balls. Then they had to stop and put their kit back together again. Pulling one off someone crying and rolling around in the dirt had a different feel to it.

The medic snipped his shirt with scissors so we could scan his midsection for blood. For shrapnel wounds, for intestines crowning out, for purple lumps or other such discoloration denoting a broken rib or a ruptured spleen. We parted the shirt halves, bracing for the worst, and gasped at what we saw.

Stretch marks. Bright and red, jagged and inflamed from the heat and pressure of his IOTV. The battle scars of rapid weight loss, which looked so much like real cuts the medic dabbed a roll of white gauze over them just to be sure.

"Grayson, where's it hurt, man?" the medic asked.

"All over," he cried.

We log rolled him, slid his pants halfway down his thighs.

"What about internal injuries?" I conjectured. If this was the case, it wasn't immediately obvious.

Dirt and specks of gravel peppered his face and hands, but nothing life-threatening.

Shining a penlight in his eyes, the medic said, "He's fucked up, for sure. Yet somehow, by the grace of God—or just dumb fucking luck—there's no entry wounds."

"Grayson, where are we, man?"

"Soccer camp," he said in a sob.

The medic and I traded looks. Under different circumstances, we probably would have found that amusing.

"Can you sit up?" the medic asked. And when he did, on his own accord, we all started to breathe a little easier.

Grayson was medevac'd for good measure. A chopper came in, and he walked himself onto it, against his wishes. He kept telling us, "Guys, I'm fine. I'm fine, guys." A slight wobble to his g's. He did seem okay. But Lord knows that an hour later or next morning even, a line of blood might appear under his nose and he would drop dead in seizure. We couldn't take that risk. Holloway hugged him before he got on the bird. Kissed him on the forehead.

"You're gonna get a Purple Heart out of this, brother." He poked him in the ribs and joked about how it might help him Stateside with the ladies.

Sans Grayson, the mood at the OP was noticeably altered. We spent the remainder of the day quiet, filling sandbags, cleaning our guns, doing what we could about our hygiene. Mentally preparing for the next attack. Would they hit us again? Yes. That was a certainty we needed to prepare for. We were so busy worrying about this fact, we failed to notice day turn to evening, evening to night. Suddenly it was twilight, and I remembered I had a cat to feed. This pleased me. I peeled open a can of Fancy Feast, tuna in gravy this time, and placed it outside the supply shack.

"Get it while it's good," I called to him.

He didn't show, so I got on all fours and started poking around. I remembered how my mom would summon our cats at nightfall. Tough-ass soldiers were brooding around the OP, eating chow and taking baby wipes to the ash on their faces, and here I was running around calling sing-song like: "Chester. Here kitty-kitty-kitty-kitty." But no one said shit about it. They all just watched me with looks of pity.

To this day, I've no idea what became of him. I like to believe that unlike us humans, Chester couldn't override his sympathetic nervous system to stand and fight with us. He was smart. He turned tail and ran for safety. Found it on some other mountain where bombs were not prone to fall from the sky. I really couldn't blame him for deserting us; we had brought something unholy to his mountain. Of course, some of the fellas had an alternative theory about what happened to him, but I prefer my version.

Holloway and the others were at the center of the OP, gazing out through optics, their heads on a swivel, a permanent state of alertness. Sound practice but also very taxing. The Mk 19 was cleaned, greased, and ready to go again at the slightest stir of anything other than friendly forces.

"Hey, bro," Holloway said to me as I approached.

"Sergeant," I replied, dutifully matching the sober energy around me.

This was usually the point in the evening when Grayson or Holloway would enrapture us with a story or two. I wanted to break the tension. Keep the tradition going. I thought about saying: "So I guess Grayson used to be fat, after all," but the spell was already broken. Grayson was en route to a hospital in Germany. Chester was gone. And no one wanted to hear any more stories.

Fiction Honorable Mention

Elaine Little

Emergency Leave

She wasn't sure how to schedule an abortion, so she googled it. Fortunately, her state was blue. The Clinic had a nice reassuring name, and it wasn't too far from her home in Chicago. She was a lucky girl, she guessed. The transportation though, what about that? She was taking a Greyhound back to Chicago. But as far as getting to the Clinic, she wasn't sure. Usually, one of her sisters needed the car for something.

As she boarded the bus, she grasped a sweat-soaked copy of all the pertinent details on a piece of paper. When she finally found a seat, she stashed her backpack and piled her stuff on the aisle seat. She didn't want anyone sitting next to her.

This was the type of gossip that would eventually get out. Besides herself, two people knew already, Staff Sergeant Metzger and the First Sergeant, so that was that. One of them would spill the beans, probably for an ostensibly "good reason." It was appealing gossip because it was girl-oriented, meaning this was not something they would have had to deal with back in the good old days before women invaded the force with their embarrassing needs.

Here went a considerable chunk of money she hadn't budgeted for. I mean, who would have thought? Her mom always said you needed to have discretionary cash on hand for unexpected emergencies. This fit the bill, and of course, she hadn't put away a cent. She still wasn't used to having money. Every payday felt a bit like Christmas.

She was a Reservist set to be deployed to Iraq within the month. But her unit was still in the premobilization phase, meaning they spent their days in a holding pattern. There were vaccinations, doctor's appointments, endless classes, days at the range where they took their time doing the tasks correctly, operating under the most optimal set of circumstances. The targets just sat there waiting to be engaged. The "civilians" in the mockup of an Iraqi town conveniently spoke English. The rules of engagement were something they memorized or read off helpful index cards.

Everything was a bit abstract. Iraq and the war seemed a world away, a world she observed only through news broadcasts, which made it seem gritty, glamorous, and exciting. The news correspondents embedded with their expensive utility vests and windblown hair.

Taking risks for sure. But looking great while doing it and with a lot more autonomy than she would get.

The bus took its time, stopping in every tiny burg between Fort Brennan and Chicago and always in a sketchy neighborhood. Carrie slumped against the window, her breath painting a round mist of fog on the window. The air conditioning was mercifully set on high, and she wished she had a sweater.

Her mind wandered. Had she packed everything she needed in her duffel bag? She'd laid it out with the others in front of the barracks. It would be gone by the time she returned, packed up on the pallets. Anything else she wanted to bring would have to fit in her rucksack. What were people saying about her?

She'd tried out her cover story on the women she roomed with in the barracks. They probably wondered why she was so chatty all of a sudden. Like, the guilty suspect going into way too much detail to the police about what they were doing the night of the crime. She had worried about what to disclose that sounded legit and worrisome, but not so alarming that they would doubt her ability to deploy. It was way out of whack for her to be going home on the weekend so soon before they were shipping out. She had told them that she had a cyst, a "female problem," and that her family doctor, who was the only one she could possibly trust with this procedure, of course, had generously offered to do it on a Saturday. It was a ridiculous story, especially the part about having a family doctor. But everyone was polite or maybe just not interested.

Carrie didn't want to tell her mother or sisters the specifics, but she wanted them to be near just in case. What if something went wrong? And what exactly did she mean by wrong? Hospital level wrong, or just I need to lie down for a couple of days wrong? She couldn't afford to look like she was incapacitated.

Carrie liked kids in the abstract. But to her sisters, it seemed kids were the goal. They conferred grace upon the owners. A blessing from God, they insisted, and then slapped their kids in the next breath. She wasn't necessarily referring to cruel slaps. These were the slaps meted out when they ran ahead of their mom into the street or put their hand out to see what was in the pot boiling on the stove. What was it with kids anyway? She'd actually seen both her niece and nephew almost follow through on such "death wishes," as if they could not comprehend that when you unbalance something, shit rains down on you. And also, in some cases, boiling water.

They didn't talk about it much, but her entire family was Catholic. Well, except maybe for Lydia, who had recently joined a Protestant Evangelical congregation, which made the Catholics seem tame.

Carrie remembered when her family still went to church as a family. Several times the congregants organized a group to meet after mass to picket the abortion clinic. Carrie had tagged along a couple of times when she was much younger. Her mom had bribed her. Promising her they'd go out to Applebees afterward. And she usually kept her promises.

They only stood by, a solemn reproach to the women's terrible deeds or as a prayerful reminder to the ones who were being led astray. It depended on who you were talking to. Carrie even liked some of the people who said these things. They helped her and were kind and encouraging about other issues. Like, do well in school, and we're praying for you. And not like praying for you in a threatening, gotcha way. Such lovely people that when they said something weird or offensive in the middle of a conversation, you wanted to unhear it. Girls referred to as whores or sluts. God as all-knowing and frankly vindictive. His unconditional love something he bestowed only upon children under the age of reason. Everyone else had to earn it, and it was easy to lose it.

Carrie knew precisely how far along she was—five weeks and two days. Well under any cutoff. The military hospital confirmed officially what she had already known. But abortions weren't permitted at military medical facilities.

When Carrie arrived home, she rushed into her shocked mom's welcoming arms. It was a nice respite before her sisters got going. As she climbed the stairs, she heard a lot of activity in their shared room. She knew her bed would be piled high with their crap, so maybe they were trying to fix that. She stopped and peeked in the door.

"Geez, girl, I thought you'd be in Iraq by now." Olga sighed as she grabbed several T.J. Maxx bags and stashed them in the closet.

Carrie flopped down on her bed. "I told you I'd call you the night before we flew out."

"Don't remember that."

It was Friday night. Her appointment was in the morning at 8. The clerk at the Clinic had said she needed to be picked up when the procedure was completed. But she hadn't figured out the logistics yet. "I've got some last-minute shopping to do tomorrow."

"Well, okay, guess there aren't too many good stores where you're at."

This was true. And Carrie's niece's birthday was next week. For a four-year-old, she had expensive taste. Or her mother did for her.

Carrie needed to pick up a gift while she was here. Or maybe she just needed to drop that. Why did she feel so obligated? Carrie's stomach rumbled. Her only symptom so far was nausea that came and went.

Her mom swept in. "You should have told us you were coming! We would have made some plans."

"There's no need. All I want is a quiet weekend at home."

From his crib in the next room, Lydia's son fussed. Olga laughed. "You've come to the wrong place."

Lydia regarded Carrie from across the room. "You look a little puffy."

"I don't think so."

She shook her head. "I don't mean fat, just . . . never mind, you look great."

Carrie clutched her stomach. "Do we have some ginger ale?"

Olga shook her head. "Nobody drinks it but you, so no."

Olga picked up her diaper bag and sighed as she pulled out used tissues, candy wrappers, and dirty pacifiers.

Carrie weighed her options. She could get to the Clinic on the bus. But the lady at the Clinic made it clear she needed a ride home. She could call a taxi. Although she didn't want the cab to come there. Maybe there was neutral territory like a CVS or something nearby the Clinic. But were there any restrictions on leaving after the procedure? Did they watch to see if you were picked up?

Her niece needed to be collected from a friend's house, so Lydia and her mom made their way out, saying they would pick up some takeout and some ginger ale.

Olga smirked as the two tramped down the stairs. "We'll get the ginger ale for you." Olga imitated her mother, not inaccurately.

Carrie dumped her backpack's contents on the bed.

Olga fixed her with a stare. "I'm not buying this 'I'm coming home one last time to say goodbye to my loving family.'"

"Which part aren't you buying? That we're not a loving family or that I wouldn't want to come home one more time?"

"I don't know, both probably? But mainly that last part."

"Why would you think that?"

"Come on." Olga liked to touch people. She advanced on Carrie and pinched her chin between her thumb and index finger. "You're breaking out a lot. What are they feeding you up there?"

"Nothing special." Carrie twisted her head.

"I know you. You are stressed out." Olga dropped her hand. "You need to tell me what's up."

Carrie ached to tell somebody, and she had a feeling her sister might be sympathetic. She could be a bitch. But at least she wasn't religious or sentimental.

Carrie walked to the window and looked out. "I'm pregnant."

If Olga was surprised, her face didn't betray her.

"And I can't have a baby right now."

Olga nodded. "That's a relief."

"A relief?" Olga often had a different take on things. But she'd never expected this.

Olga threw her arms out in a helpless gesture. "Can you imagine all three of us in this cracker box house with three kids?" She eyed Carrie's stomach. "You're not showing a bit. But aren't you off the deployment now? So welcome home, I guess?"

"If I take care of this now, they said they'll let me go. Supposedly I'm getting a cyst removed."

"When is this abortion thing happening?"

"I have to be there tomorrow at 8."

"You need someone to take you probably."

Carrie ran through the possibilities in her head. "Well, not necessarily. Maybe I can take a bus there and. . ."

"Stop. Just stop it. I can take you."

Carrie was relieved. "Can you just promise not to tell anyone?"

"You mean like Lydia and Mom? Like that would be a good idea." Olga walked to the door. "Hey, I gotta go out for a few minutes. Tell Mom to save me some food." She hesitated a moment before stepping into the hallway. "Be ready tomorrow at 7. We're stopping at Dunkin Donuts on the way, and you're paying."

After the loud voices and insistent questions of her family, it was a relief to lie down for a minute in the quiet. After a while, Carrie realized she was hungry, so she headed to the kitchen. She wasn't willing to wait for takeout. It had been an exhausting day.

Afterward, she lay on the bed. Olga hadn't exactly said she wouldn't tell anyone. If so, it wouldn't be the first time her mom was subjected to news she was better off not knowing. Still, her being deployed to Iraq meant everything would have blown over by the time she got back.

She checked her phone. No texts, no messages. Her friends all thought she was already overseas. She got up and switched off the light and turned her face to the wall. With luck, she'd sleep soundly and wouldn't wake up when her sisters returned. But she knew there was no guarantee they'd be using their inside voices.

When they finally returned, her mom and Lydia's concerned murmurs were actually more annoying than Olga's matter-of-fact declarations. Apparently, they had all been briefed about the "cyst" and were predictably freaking out.

Her mom whispered that this was ridiculous; a cyst should have been enough to keep her home. Carrie almost got up to explain. But the two Tylenols she'd downed with a beer chaser were enough to clamp her drowsily to the bed. As the wooziness overcame her, the conversations grew distant, like hearing voices through a glass pressed to her ear against a door.

At 6 a.m., she awoke. Lydia was up with her daughter. Olga was snoring away. Carrie pulled out her laptop to check out the route. She was anxious. The pill option was out since it would require too many visits. She'd be getting an aspiration abortion, which was more invasive. It was safe, she knew. Still, her hands were clammy.

Olga's alarm went off at 6:55. She eased out of bed in her sweatpants and T-shirt and glanced over at Carrie on her laptop. "Alright, let's go."

Carrie shut off her computer. "Aren't you going to get dressed?"

"This is dressed. I'm just going to check on Ava."

Ten minutes later, they were in the car, scrutinizing the printed-out MapQuest directions.

Olga sat in the driver's seat. "We don't need that. I know this place."

Carrie crumpled up the paper.

In the distance, a Dunkin Donuts loomed. Olga sped up and made a sharp turn into the parking lot. Carrie gripped the seat. They parked, and Carrie passed her a wad of bills.

Ten minutes later, Olga walked out carrying two coffees and a bag stuffed full of doughnuts, breakfast sandwiches, and bottled juices. Carrie opened the door.

Olga grinned as she handed the bag to Carrie. "I spent all your money."

Carrie poked her nose into the bag. A few coins jingled at the bottom.

Olga gulped her coffee as she maneuvered out of the parking lot with one hand on the steering wheel.

Carrie steadied her coffee. "So how do you know this place? Is it the place Mom used to take us to picket?"

Olga nodded. "Yes. I'm surprised you even remember that. You were such a little thing

"Do they still do that?"

"Yeah, they usually got a few people out. They're pretty chill, though. Just holding those signs and looking depressed."

"But nobody we might know, right?"

Olga shook her head. "The new priest, he's not into that. Now, Lydia's new church, maybe."

"Is Lydia still Catholic?"

"Don't think so. Just the other day, she lectured me about how I needed to accept the truth and come visit her evangelical church."

Carrie smiled. "She ought to know better than that."

"Shit!" Olga stopped at the train crossing as the gates descended, almost hitting the car's hood. The horn honked loudly as the train approached. She peered to her left as she backed up. "Ugh, it's a freight train too."

The cars crept by in slow motion. Carrie checked her watch, worrying about whether she would make her appointment. Olga, who was not good at silence, kept fidgeting. "You know why I'm helping you, don't you?"

Uh-oh, Carrie thought. Did she want something besides a bag of free doughnuts?

"I know I told you that I thought the recruiter in our high school was a con man and that I was too smart to get pulled into his lies. But I'm here to tell you, I almost joined."

"What happened?"

"Ava happened."

Olga's fingers drummed loudly and insistently as she watched the train cars lumbering by. "I just wanted you to know that I don't still think you're stupid for joining."

For a while, neither spoke. They just stared as the cars rumbled by.

"So, you've been at the clinic recently?" Carrie finally said as they drove across the tracks.

"Yes, it's a cheap place to get checkups, medication, and . . . I may as well tell you, I got an abortion there once."

Carrie almost spit out her coffee. "Does Mom know?"

"Nobody knows, well, you know now."

"When was this?"

"Sophomore year. And I still feel shitty about it, I really do. And then I got pregnant again, and believe me, if I wasn't so mega fertile, this shit would not be happening. But I know Ava is the same child as the one that got aborted."

"What?"

"Shit, girl. Listen to me. I'm saying she's the same kid. The same spirit inside. I feel it in my heart. That's enough for me. Maybe not enough for you, though. You think too much."

"Well, God works in mysterious ways."

Olga made a face. "There you go making fun of me."

"I'm not making fun of you."

Olga waved her hand in dismissal. They pulled up to the Clinic.

"Alright, you're here."

Carrie grabbed her coffee and picked a doughnut out of the bag. She backed out of the car.

Olga revved the engine.

"Wait, aren't you coming in?"

Olga looked over at a small group of people gathered in front of the entrance holding signs and passing out pamphlets.

"No, thanks. I ain't got time for this. Give me a call when you're done." The car roared off.

Carrie sipped her coffee as she approached the protesters. They tried to press pamphlets into her hand. Then a clinic worker walked out, took her arm, and escorted her the rest of the way.

Fiction Honorable Mention

Kristine Otero

We Go Together

The quarterback for the high school team. A celebrity in his own right.
Small towns breed poverty and patriots in equal proportions.
He joined the fight to win the war on terrorism.
Convinced his touchdown would even the field. Jesus on his team.
A born soldier—a leader. The monotony his calling. A sergeant at twenty-two.
Living the Army values, he could call a 9-line in his sleep.
Three tours and a million buddies, at twenty-three he understood Farsi.
The layout of the country more familiar than home. He knew every rock and pothole.
Each deployment, the smells of rot became a little more comforting in their predictability.
His armor kept a secret, his leathery neck told the story of his war.
Just another day in paradise, where the livin's easy.

When the firing started, he was calm—*push through.*
Saving two lives already, a hero in his own right.
Hammering up ahead, he drove into it, like he was trained to do.
A graze on his shoulder, the bullet found shelter in his gunner's femoral artery.
Get through it. He radioed—to nowhere.
Doc was the Gunner—the gunner named Gunner, his best friend.
Collapsed in the humvee, blood spreading like fire—the smell of metal and gunpowder.
Dragging Doc Gunner out, belt in his hand. Cutting off his shirt with a pocketknife.
Each move fast and sure. Cut his BDUs—rip—tie the belt—staunch the blood with his shirt.
30 seconds—done. "DOC," he yelled, over the pop and rat-tat-tat of fire.
Doc was bleeding out on Highway one. A hero in his own right.
"We go together, or not at all," Doc always said.

26 days—25 sleeps, that's all they had left in country.
He turned around one last time, looking at the "home" he'd never see again.

Boarding the aircraft with Doc—*we go together or not at all.*
Casket tied down in the uninhabited C130. A frigid, empty Angel Flight.
Lulled to sleep by the roar of engines, and the cold, he slept next to Doc all the way home.
We go together or not at all.
"If you love what you do, you'll never work a day in your life," Doc once said.
And so it goes—Doc never worked a day in his life. It was all he could say to Doc's mom.

Five years, seven months, 23 days short of three tours. He wouldn't reenlist.
He was never a lifer, but he didn't love his job anymore.
Quiet—he wanted quiet. Craving home, a job—where the livin's easy.
Going home with all of his shit was different than leave. Finality felt oppressive.
His parents hugged him, stifling heaviness filled the air.
Disappointment hung in the space between them, Mom's smile never reached her eyes.
Dad exhaled the question, "Whatcha gonna do now?" and held his breath.
Twenty-six, going on twelve, he asked if he could stay and rest awhile.
Mom and Dad were patriots, kind conservative Christians, never climbing out of poverty.
Reliant on the eldest son, a possible turncoat, who Jesus seemed to have abandoned.

Waking to the smell of bacon brought fear where it used to bring joy of the temporary.
He had slept for almost two full days, allowing his circadian rhythm to reboot to the States.
Fresh start—priorities. The black shine of the used truck forced a smile of pride.
The sole salesman and owner asked, "You're the kid, the war hero, right?"
Stepping back, looking to his feet—ashamed. "I served, Sir," he said.
"I'm home now."
A slap on the shoulder, "If you need a job, you have one here, kid," grinning wide.

He could throw a ball and aim a rifle. He could run a touchdown and a convoy. He would learn.

Country nights were quiet, his screams echoed. He only spoke in nightmares now.
"Let me know if I can help" hurt more than the tingle of the bullet that passed through his arm.
Safety in the compound of cars, compromised by the rotating shoppers.
The glare off the glass and asphalt, endless entry points, required smiles. He'd never learn.
Cashing his only paycheck, handing it to his mom, minus a twenty. She stared around him now.
Maybe in the periphery her son still existed.

He had only been to the bar as a boy, with his dad to watch games. He slipped in—a man.
One finger raised to the bartender, the universal sign for a beer. His twenty on the bar.
A game was on, he pretended to watch, the screams in his head getting louder.
Pool balls clanked. Glasses clinked. Monotone announcers and "oohs" from the crowd.
An overwhelming cacophony. The smell of sweat and stale beer. Blood pumped in his ears.
Pocketing the change—pushing the vomit back down, rising slowly from the stool.
He felt the imagined weight of his gear, the pressure—a comfort, an illusion of protection.

He applied for the night shift as a security guard, a sugar plant on the edge of town.
"You're the war hero," said the foreman, tossing his application. "You've got it."
He bought a gun to protect the sugar plant. A sigh of safety and purpose.
Alone, preferring to walk. The brisk air and stink of beets, reminiscent of "home."
A stray dog walked beside him each night, vigilant. At 0300 they would rest awhile.
"We go together or not at all," he would tell the dog.

He could feel the warmth behind his eyelids, while a cold bit under the weight of his gear.
Struggling to awaken from his deathbed dreams. He peered into the harsh morning sun.
The pain, hot and throbbing—every breath a gasp. He tilted his head to Doc.
Reaching, he squeezed a sticky hand. Fire spread across his chest, his limbs frozen.
War is the long farewell for the short goodbye.
We go together or not at all.

J.G.P. MacAdam

Real Red Blood

Let's play a game. I name a character, a scene, a quote or whatever and you name the war movie. Okay, ready—go. I am in a world of shit That one was a gimme. I love the smell of napalm in the morning. Okay, also a gimme. Demi Moore with a buzzed noggin doing crunches like a fucking champ. "I wanna go home, Forrest." Orlando Bloom falling ass-backwards out of a Blackhawk. "My full report, General." Anything over the River Kwai. "Frrree-dom!" A couple of ace pilots shirtless and glistening with oil, playing some epic volleyball.

All I want is to watch war movies again.

They're gonna take my fucking leg.

Ugh, I gotta stop. Even thinking about war movies makes me queasy. It's been years since I watched a war movie. Every time I watch a war movie I throw up, and I don't mean hiccup my dinner into the back of my throat, I mean projectile vomit, blowing chunks, ohmigod-what-is-happening-to-me—am I dying?—throwing up. I'll be down on the floor clutching my stomach, dry heaving, tears running down my cheeks, till I finally turn it off.

I don't know what's wrong with me.

Thing is though, I love war movies. I used to watch war movies for breakfast, for dinner, on my phone, laptop, in my apartment with the sixty-inch. *Band of Brothers*, anyone? Holy shit, nobody watched *Band of Brothers* more than me. I used to literally recite every line verbatim. But I can't anymore. Even spoofs like *Whiskey Tango Foxtrot*, *Three Kings*, or *MASH* roil my guts to the point where I spend the rest of the night in a cold sweat.

Except tonight is the night when everything changes.

I'm gonna watch a fucking war movie, so help me. I'm not gonna let this sensitive-stomach-bullshit beat me. I got my movie ordered right here in front of me. *Atonement*. Okay, maybe not quite a war movie, or even a military movie, but it's got Dunkirk and other WWII stuff in it. It's the least war-y war movie I could find. A baby step in the right direction. A slow acclimatization of my stomach back to the shell-shocking, corn-syrup-blood-squirting sagas of fucking patriotism I grew up on.

Maybe I should get a bucket, just in case.

How the fuck did this sensitive-stomach-whatever-it-is start? I can't remember. That ever happen to you? Like, something happens to you and you can't remember how it happened? But then it hits you in the next second, like it just hit me, what you were trying to remember?

Yeah, that's how it started.

At that fucking crip camp.

It was some disabled veterans retreat or some shit. Up in the hills. Scenic. I forget what they actually called it. You had your spinals, your burned guys, your missing-limb guys, then guys like me getting around with a fucking walking stick. Everybody in little groups talking about their memories and experiences, their feelings. I fucking hated it. There were like two girls in the whole congregation. Two girls, that's it. All I wanted was to get laid. That's the only reason I signed up. Fat chance of that happening though. Should've known it was gonna be a fucking sausage-fest.

So, I drowned my sorrow by driving my fat ass down to the 7-Eleven, buying some whiskey, driving back up, smashing half the whiskey in the parking lot, then ambling over to the epic fucking buffet. Mostly I just wanted the red stuff. The raspberry syrup over my vanilla ice cream. The prime rib. The watermelon. The red gelatin salad. Ohmigod, I fucking fell in love with that red gelatin salad, just watching it jiggle on the plate. Jesus, I was fucking drunk. Then this one guy—Ryan, I think his name was—comes wheeling up to me, a healthy sheen of sweat on his forehead from playing b-ball with his crip buds.

"Hey, man."

I already don't like his face, too movie-star.

"Hey."

"I saw ya sitting over here all alone and figured I just come say hey."

Another mouthful of gelatin. "Hey."

"Man, that's a sweet cane, or staff, you got there. You carve that yourself?"

"No."

"Is that an Indian etched on there?"

"It's an Apache warrior."

"Sweet. That like a heritage thing?"

"No."

"Okay. Well, I thought I'd just spin by and see if you wanted to come play some b-ball. You might be taller than us, but we'll go easy on ya." Wink. "Or maybe later, just you and me, we can shoot some hoops?"

"Do I look like I play basketball?"

"Well, I—uh, you do look like you could use some exercise, to be honest. But, hey, we're all friends up here. We all got, y'know, shit we're trying to work out. That's why we put this whole thing together. So we can work on those things. As a family. As a team."

Oh yeah—I recognized him now. He was one of the volunteer camp counselors.

"Nah, man. I'm good." Another mouthful of gelatin. "I'm jush gonna 'it here and ea' m'pood."

"Sure thing, man. Just let me know if you need anything, alright? Don't be a stranger!" He spun around and vrrr-ed down the sidewalk. One of the only two girls wheeled up beside him. I could tell they were talking about me. The way he thumbed behind him. The way she glanced at me. Was that disgust in her eyes? Maybe I was just projecting. God knows she probably just wanted to be treated like every other broken-ass vet up there and *not* eye-fucked everywhere she went. Wouldn't that be showing some team spirit? Whatever the fucking case may have been, I just got up and limped my own broke-ass over to the buffet for some more gelatin.

It was the following weekend, when I went to go see *American Sniper*, that it happened. I'd been looking forward to seeing *American Sniper* for like a fucking year. It was a hit. People were lined up around the block like it was the fucking draft. I couldn't even make it past the opening scene. He's about to snipe a kid and the kid's mother, before either of them can lob a grenade at a convoy of friendlies. We watch his finger rest on the trigger, his eye through the other end of the scope, his fucking eye, he breathes out, here it comes, they're gonna die, their lives are gonna snuff out with just a couple twitches of his forefinger, it's his burden to bear though—isn't it?—his eye we see through, not theirs. Then there's me practically running across people's laps to get to the bathroom before my stomach turns inside out. Haven't been able to watch a war movie since.

Bucket in place.

Commence countdown.

I'm gonna kick this. It's now or never. They're gonna take my fucking leg. Don't think about that. Gotta concentrate. I hit play. Opening music, credits. I am not thinking about vomming. Who's thinking about vomming? This is just a nice romance movie with some war in it. Don't think about vomming. Funny how WWII and romance go together all of the time. I mean, turn on BBC, CBC or PB-fucking-S and there's always this Brit drama about a girl and a guy being separated during The War. She looks off somewhere, thinking of him. He's writing a letter, thinking of her. Maybe he won't make it back, maybe he will, then he finally does and they live happily ever after, sorta, kinda, maybe, however it works out. That was the script, the formula, the way it was supposed to happen. Guy goes off, girl waits at home. Romance and WWII, going together like vodka and Red Bull. That's why so many people have this like love affair with WWII. America finally got over its inferiority complex with Europe and took its rightful place—some would say—as top gun. It was the fucking good ol' days when shit actually followed the script. The script: get attacked, declare war, young men sign up in droves to go fight for their country. It's a defunct script anymore. Wars nowadays don't follow the script. Wars don't end in victory, they just piddle along like an old man's piss stream. Still the script's what people expect, what people want to see. People paid good money for this war, they wanna get their money's worth, amiright? I want WWII—the sequel! That's the fucking standard. I want Americans greeted as liberators and showered with praise and flowers and pussy and all that shit. Roll credits. Everybody goes home happy, buys a house, has kids, eats a TV dinner.

But mention the War in the Pacific and people be like, Wait, I thought we were talking about WWII?

Anyways, I'm not really watching *Atonement*. The actors are humping in a library or something. My stomach gurgles. This isn't going to work. Maybe I should just watch cartoons.

I mean, it never goes the way it's supposed to go, amiright? In the movies, the plan has to get all fucked up. That's where the drama comes from, the suspense. Though, it's all part of the script, right? Like when Maximus tries to escape and rejoin his army and march back to Rome to take on the Emperor, but then the Emperor captures him and drags him back to the Colosseum—that was the plan getting all fucked up right there. But it's still part of the script, the overall plan, the way the story's supposed to be told, you see?

Take any Vietnam movie. The whole war was an exercise in things not going according to plan. But that's what the post-Vietnam-era war movies did, they took the old script with the flawless hero, the cereal box hero—the John Wayne from *The Longest Day* and *Back to Bataan* and whatnot—and they made him into these flawed heroes, these tragic heroes, these psycho ex-military mass murderers (Rambo, anyone?). But it's still the *American* hero taking center-stage, mind you, which means he was a he and usually white. I mean, who wants to be the Vietnamese in those movies, amiright? All of the no-name Vietnamese extras screaming and fighting and dying and nobody understands what they're saying, if they say anything at all. That's the war, that's any war. Endless nameless extras dying by the napalm-scorching bushel. Don't even get an end credit. That's how Iraq became Vietnam, or Vietnam Iraq, I don't fucking know. People were already setup for the tragic war story. The quagmire. The why-the-fuck-are-we-here war.

Oh yeah, Afghanistan. I forgot about Afghanistan, and Korea, and Grenada, and Serbia, and I'm sure there are other wars, operations, conflicts, whatever, but I can't remember them right now. They just don't have the scale, the oomph, the attention, the national narrative, the box office hit potential, do they? We still even in those countries?

They're gonna take my fucking leg.

The spherical ice cube in my vodka and Red Bull is an eyeball. There is no eyeball. I'm seeing shit. I should've fasted before watching this fucking movie. But Percocet and alcohol just kinda go together like memory and imagination, amiright?

That ever happen to you? Like, you remember something and you're not sure if what you're remembering is something that happened or something that you imagined happening? Like, growing up on war movies, I imagined so many combat scenarios, like I imagined Iraq before it was even Iraq for me, you know? Now, afterwards, I'm like did I imagine fighting in trenches? Cuz there were no trenches in Iraq, right? Or was it a canal? I mean, it's been *years* since I was in Iraq, maybe I just watched too many war movies and all the wars and all the memories and all the stupid shit and nonsense shit and the well-that-makes-fucking-sense shit all just kinda blended together.

Like what happened to me last night.

It happens sometimes. I wake up just seeing that fucking eyeball. I don't know if the eyeball ever even fucking existed. It must have. It definitely did exist. I don't know. My buddy, Elias, he visited me not too long ago and he was like, "Remember that fucking eyeball you found?" So if Elias's memory validates my memory, it really fucking happened—right?

Anyways, it was hot. Like roasting. Like you're-in-Iraq-and-it's-the-middle-of-the-afternoon-in-fucking-August hot. Some really nice news came over the squawk box about the explosion we just heard a few clicks over. It was a suicide bomber. Hit a market. Dead bodies everywhere. We were an infantry platoon—Apache Co. boyz, holla!—but we ended up tasked with heading over there and cordoning off the area. Cool, get to move around after atrophying in the same fucking spot for two fucking days. But then we get there and the squawk box says, Okay, we're gonna help clean up the bodies. What? Ah, ya gotta be fuckin' kidding me. But we ante-up and get the job done. There's blood everywhere. I mean, real fucking blood. Unlike the movies, real red blood isn't red. Real fucking blood is all kinds of other colors, it's every color except red. No, I'm not colorblind, I'm just telling you that real red blood is everything *but* red. It's like rust, like dirty oil, black when soaked in a hijab, it's brown, it's purple, it's even fucking blue. It was in broad splats across the pavement, already dried. It was caked goop on the pieces of person-meat we picked up and stuck in these little see-through biohazard bags. They at least gave us latex gloves. God, there were meat pieces everywhere. Big smoky black blast patterns in a semicircle, debris tossed, fucking tomatoes, and that's when, as I was picking up a couple of meat chunks of this lady, I found her eyeball. It was whole, like just the optic nerve or whatever hanging onto the back. Its iris was brown and perfectly circular. Looked like you might just pop her eye back into place and it'd be fine. Little dirt on it, on one side. Just blow it off. The sliver of meat on it was cooked purple under the sun.

I grabbed her eye and baggied it up.

Then my platoon sergeant, Sergeant Barnes, hollered at me to go do something else, Sergeant Barnes who was dead, he fucking died, as in—boom!—pieces of him and his Humvee all over the place. That was like a month later though, I think. He was dead before we found him, deader than dead, man fucking died, I said, as in buried six feet under in the dirt in some north Kansas cemetery, the prairie wind whistling over his tombstone, drying his widow's tears before they can even leave her eyes.

I think I imagined that last part.

Anyways, the eyeball, without anywhere to immediately put it, the collection point for all the meat pieces being in one direction, Sergeant Barnes hollering at me in the other, I put the baggied eyeball in my pocket. I'll drop it in the meat bucket in a second after I see what's up with Sergeant Barnes. Then we mount up and go back to the spot we were in before, staring at the endless horizon, at nothing, thinking about all the war movies I could be watching right now. I had been on a classics binge, at the time. I think I was somewhere like halfway through *Le Grande Illusion*, a French film about WWI, boring as fuck, by the way, barely any action in it, check out *Paths of Glory* instead. Any-fucking-ways we ended up out there guarding an empty patch of sand for like another two days. Another platoon even came out and resupplied us with water and MRE's we were out there so fucking long. Talk about great un-expectations. We were *way* off-script. So, finally, we head back to base, all I want is a shower and some hot chow, but I never get what I want. We gotta cap off the trucks, clean our weapons, make sure of this and that before any of the kickback and take-a-night-off shit can happen. But eventually, finally, I take my pants off—bottoms, they call them bottoms in the Army. I take my bottoms off and hang them in my closet thing I got. I've got bottoms for all kinds of different duties. I've got my missions bottoms, the most comfy but the most worn out; my guard bottoms, for those days when we're just jacking it in the guard towers on base; I got my heading-the-fuck-home bottoms, which were squeaky clean and never worn until that final day when we boarded the bird home. I can see them there, all hung up dress right dress in my closet thing.

My bunkmate, Elias, and I go to sleep.

Next day, we've got another fucking mission. Fuck me, I didn't even get a chance to wash my mission pants—bottoms, whatever. I open the door to my closet thing and ohmifuckingod what the fuck is that smell?

"Jesus, what *is* that?" says Elias.

I pull out my pants and there it is, in the pocket, the fucking eyeball.

It wasn't red anymore, I'll tell ya that.

Finally we're at the part in the film where there's gonna be some fucking war. There's a sandbag bunker in a London street. Johnnies in their uniforms. Any minute now it'll skip to France or wherever and it'll be game on. But I can already tell my stomach's not taking it. This was a bad idea. I don't want to watch anymore. C'mon! You can hack it! Suck it up! But I don't want to. Before even a bullet or a piece of shrapnel can fucking fly, I empty my guts into the bucket. I'm crying, begging, praying for it to stop. It's the kind of puking that hurts. My head's on fire. The muscles in my mouth fucking ache. My nostrils sting with stomach acid. I see one drop of blood on the screen and it's that telltale corn-syrup red, and I'm on my knees retching so bad I can't even fucking breathe. I fumble for the remote and turn the shit off.

They're gonna take my fucking leg.

That's what I say to the physical therapist.

"Sometimes, physical therapy can only take you so far," she says. "Everything's still pretty shattered in there."

She's telling me the same thing the doctor did: all the plates, pins and rods in the world can't put Humpty back together again.

"But you're slated to get a new leg, right?" she says. "A prosthetic?"

"Yeah, they're already like measuring where to cut and shit based on the type they wanna give me."

"I've seen this sort of thing before. It really is the best decision. There's just nothing more to do with it, except replace it."

"I know."

"Otherwise—"

I already know otherwise. Otherwise, I'm hooked on pain meds for the rest of my life. Otherwise, they start to fuck with my head the way pain meds do. Otherwise, I end up a shell of the person I once was, or am, I don't fucking know. But the thing is, I'm already a shell. A big fat staggering shell of who I once was. Nobody has the courtesy to call me a fat fuck to my face, but I'd call me a fat fuck. This wasn't part of the script. This wasn't how it was supposed to go. I never saw myself ending up like this.

"How're you feeling about it?" she says. "Even with the chances of a prosthetic really working out better for you, still . . . it's gotta be a little nerve-wracking, about the surgery, I mean. About losing a part of yourself you've been working so long and hard to try and keep."

"It's just meat."

The next morning I wake up and I swear I'm staring at that fucking eyeball. She saw her whole life through that eye. One of two, most likely. What had she seen with it? She see any war movies? American-made ones? She see the fall of a dictator? She see her children born? She see her children die? She see the suicide bomber coming, she see it was too late to run, she see her life flash before her eyes?

I can only imagine, my imagination's gotta fill in the gaps where memory won't serve. Unreliable, memory. But war movies, the so-called great ones, they last a long time.

It's not an eyeball. There is no eyeball. It's just the cap to my empty vodka bottle. Just another nameless extra, meat on the street. There's nothing to see, anymore.

I try to get up and hit my head on the corner of the coffee table. "Ow! Fuck." I pop a Percocet and hobble over to the sink for some water. I swallow and stare out at the morning. Another morning. There is no escaping the script, the one I've written for myself and watched play out time and time again. That's the hardest part, I realize, rewriting the script.

Something warm drips down the back of my neck. "The fuck?" I reach back, touch the tender spot on my scalp, pull my palm forward. "Huh, real red blood."

Peter Trivelas

The Drive Home

There were stories. There were types of anxiety on board. You couldn't go ashore if your gonorrhea wasn't cured.

The ship was making its way back from the South China Sea to Virginia's Chesapeake Bay—cruising at 30 knots, 24/7. We crossed the equator. Twice. Heading south, then north. It looks exotic on paper, but it was all water. It'd be a long month of nothing but water.

Riding low, like a car with a trunkful of contraband. The scuttlebutt, and I think this is true, was that the souvenirs everyone had bought were to blame. The low prices for electronics in Asia were irresistible: Pentax cameras, Nikon lenses, Sanyo boom boxes, Sony reel-to-reel tape decks, enormous Pioneer speakers. Months of shopping, 3000 sailors, made for a significant load.

Families anticipated our arrival. We'd be standing in a neat line along the edge of the flight deck in our dress blues, at parade rest, legs slightly apart, hands locked behind our back as the ship neared the pier. You'd never know from seeing us from afar how badly we'd all wanted to cross that plank to shore, how we could barely stand upright.

Hundreds of girlfriends, wives, and kids would be onshore waving flags and holding up welcome signs like *Make Love, Not War* on huge banners. Nailed it. It was on everyone's mind.

Definitely on mine.

Many of the men journeying home were healing from the clap. Syringes emptied into buttcheeks, four inches of needle and antibiotic.

Limping sailors with sore butts were a familiar sight for a week after any of our ports-of-call. It took two weeks for the medicine to work its magic. The voyage home would allow plenty of time for most cases. Most, but not all. My eyes were still not back to the previous 20/20 they'd been all my life after my tryst with Sakura.

The route to our home port in Virginia took us east across the Pacific Ocean and around the southern tip of Argentina. Cape Horn is the sea lane between South America and Antarctica, the meeting place of the Atlantic and Pacific Oceans, a famous nautical passage notorious for its dangerous combination of strong winds, huge waves, wild currents, and even icebergs, depending on the season. Glad I didn't know that beforehand.

The first two weeks traversing the Pacific were idyllic. Most of the men were done with wartime duties. Flight deck became a sand-less beach littered with spent sailors sunbathing on blankets in between the A-1 and A-4 attack planes securely tied down. Recently purchased sound systems taken out of boxes and tested. Warriors reading, dancing, and day-sleeping to make the best of the monotony. Recounting drunken nights, bragging about port romances. Laughter. The recovery process, decompressing.

I was propositioned more than once on that drive home—"Haven't you ever heard of male love?" being my favorite. Even stumbled upon an intimacy in progress in a storage room. I'd later call bullshit on The Village People's denying their song *In the Navy* was about sex.

Vacation conditions ever-so-gradually morphed, ship rolling and lurching with building intensity as we approached the Cape. It was enough to make a man turn green, which I did. Cold, overcast air materialized. Sailors in color-coded shirts rolled fifty jets off the flight deck, one at a time, onto elevators and crammed them securely into the hangar bay like puzzle pieces, one deck below.

The flight deck was fifty-five feet above the water line, which meant the waves crashing over the bow onto the now-barren deck were seventy, eighty, ninety feet high. For one risky moment, I ventured to the middle of the flight deck, a lone figure in the fierce wind, up in the face of mortality, surfing on the world's largest surfboard, longer than two football fields. Legs planted, arms out, keeping balance, defiantly in full view of the ship's captain on the bridge. It was a break in the tedium, a foolhardy move only depressive boredom could birth.

As we sailed on, the carrier was thrashed like a forty-foot sailboat in a mean squall. The hull slapped the waves, twisted and moaned like it might break apart and be swallowed by a roiling monster. Fear, seasickness rampant. With freedom only a week away, the ironic possibility that we could become yet another Cape Horn shipwreck intruded into my happy anticipation.

It passed.

Time for the final pee tests—for some, a far more precarious storm.

The hundreds of men who'd reported to sick bay for penicillin shots three weeks before were once again standing in hour-long lines, this time to find out if they were clear.

Word went round that there once was a guy who wasn't one hundred percent sure all his symptoms had disappeared. His wife and kids would be waiting on the dock to welcome him home, so he asked his healthy friend to pee in the cup for him just to be safe. He needed this insurance policy. His friend's pee tested positive. He gambled and lost.

How he explained that to his family is a mystery. Reckless infidelities coming home to roost.

We entered the final stretch, ghost ship heading up the East Coast of both Americas. Lingering were memories of pilots who hadn't returned from their missions, men who'd jumped overboard, and tons of bombs dropped from jet wings.

Anticipation of freedom and happiness prevailed—neither of which would ultimately match the impending reality lying in wait. But only those of us who'd walked the plank before were aware of that.

David Blome

At the Curtains

Ed stopped at the back door and sighed. He slid the key into the lock and felt his hand start to tremble. Inside, he placed his satchel on the floor and relocked the door. Now his heart was racing. Swallowing hard, he turned and walked the twenty-three steps through the kitchen and living room to the front window. He stood there for about three minutes. Then he took hold of the curtains, closed his eyes, and pulled. The sunlight struck his face. He took a deep breath and opened his eyes. Nothing.

"Thank God," he whispered and closed the curtains.

*

In the back of the church, Ed stared at the stained glass and waited for Father Kevin to finish his conversation. When he did, Ed walked straight towards him.

"Ed, it's good to see you."

Ed smiled. "Thank you, Father. How are you?"

"Great. How's your son?"

"He's doing well. I spoke with him last week."

"You did? How often do you talk?"

"About once a week, when he has a chance. We usually talk for a half hour or so."

"Does he have to pay to call?"

"He does."

"That must be expensive."

"Actually, Ben says it's cheap."

"All the way from Iraq?"

Ed shrugged. "That's what he says. I'm not complaining."

Father Kevin laughed. "Well, it must be great to hear his voice."

"It sure is."

"Next time you talk to him, tell him I'm praying for him."

"I will, Father. He'll appreciate that."

Father Kevin squeezed Ed's shoulder. "Will I see you tomorrow?"

"Of course."

Ed left the church and walked to his car, grateful for the brief conversation. He turned the key in the ignition and checked his watch.

"Five past eight. Perfect."

About twenty minutes later he arrived at work to an empty parking lot. Inside, he unlocked his office, hung up his satchel, and turned on the computer. This was Ed's favorite part of the workday. For the next half hour or so he had the building to himself.

While his computer came to life, Ed stared at the picture of Ben on his desk. The previous year, the two had gone deep sea fishing off the Outer Banks. Ben caught a thirty-four-pound bluefin, and Ed immortalized the event with his son's camera. After the trip, Ben printed and framed the picture and gave it to his dad as a birthday gift. It was the only picture in the office.

Ed reviewed the day's agenda and opened his email. A new message from "Dwyer Sgt Benjamin" brought him to the edge of his chair: *Dad, The package of books you sent made it. Can't thank you enough and can't wait to dig in. Reading has been the best way to pass the time. Sorry I couldn't call today, something came up. I'll try later in the week or next Sunday around 1100 your time. Hope you're enjoying the weather back home. It's still hot as you-know-what over here. Love, Ben.*

Ed checked the time Ben sent the email. "Only a few hours ago," he said. Then he reread his son's message. He pictured Ben opening the package. Ed had worried all afternoon the day before because Ben didn't call. Now he felt silly. "Of course something came up. He's not working a nine-to-five. Get a grip on yourself, Ed." Ed said all of this out loud.

He then gathered his thoughts, hit Reply, and wrote, *Dear Ben, I am very glad that you received the books and I hope you enjoy them. I will look forward to talking with you. Take care of yourself. Love always, Dad.* Ed stared at the email. I'll write more later, he thought. He hit Send.

Just then his secretary entered the building. She walked straight to Ed's office.

"Good morning, Ed, how are you?"

"I'm good, Tonya. How was your weekend?"

Tonya relaxed. Smiling, she said, "Busy, but no complaints. How was yours?"

"Uneventful."

"Just the way you like it."

"You know me."

The rest of Ed's Monday passed much like any other. At the end of the day Tonya stood outside of Ed's office and waited for him to look up.

"I'm about to leave. Do you need anything else from me?" she said.

"Not at all. Thank you for your help today."

"Just doing my job."

"Right," Ed said with a smile. "Have a good evening."

When Tonya left, Ed walked to the assistant director's office. The door was open but he knocked on it anyway.

"How's it going, Jim?"

"It's going, Ed. What's up?"

"Not too much. Are we all set for tomorrow?"

"Yup."

"Great."

"Hey, Ed?"

"Yeah?"

"You heard from Ben?"

Ed scratched his cheek. "I did. He's gonna call later in the week."

"Oh, good. Does he need anything over there?"

"Just books."

"Just like his old man."

Ed smiled. "Mm-hmm."

Before leaving the building, Ed sat at his desk and wrote an agenda for the next day. Satisfied with everything, he shut off the lights, picked up his satchel, and walked to his car, the only one left in the parking lot. As was his custom, he drove home in silence.

Had Ed driven straight home, he would have walked through his front door about twenty-five minutes from the time he left the office. But Ed had not driven straight home in months. Instead, he drove well out of his way to enter the neighborhood via its western entrance so he could park on Washington Drive, the farthest street from his house. When he took the keys out of the ignition and stepped out of the car, he had to walk another seven minutes down two residential streets and across three yards to arrive at his back door.

Standing there, he sighed, slid the key into the lock, and felt his hand start to tremble. Inside, Ed again performed his ritual. When he passed through the living room, where he had once played and wrestled with his son, he cleared his throat. Three minutes later, curtains in hand, he shut his eyes, took a breath, and pulled. He looked. Nothing.

"Thank God," he whispered and closed the curtains.

Ed knew that if they had to come, they had to wait. It took Ed an extra sixteen minutes to drive to the other side of the neighborhood, arrive at the back door, walk to the front window, and open the curtains. He wouldn't know they were there until he opened his eyes. If they were, that is, if Ed really had lost everything, his ritual at the curtains gave him an extra sixteen minutes to think that he still had a son.

*

Inside the tent, Ben used his poncho liner as a curtain to block the entrance to his living area. He had to. His space had become the platoon library and he had no shortage of visitors.

Berks, one of his best friends, pulled the poncho liner away, grinning. He cleared his throat and shouted, "Hey! Wanna get some chow? Ben!"

"What?"

"Wanna get some chow?"

"What time is it?"

"I don't know, like 1800. You sleeping?"

Ben looked at his watch, squinted, and rubbed his eyes. "I was."

"Sorry, couldn't tell. Your eyes were closed. Wanna get some chow?"

"Yeah. Give me a minute, asshole."

Ben swung his feet out of the cot, put on his boots, fastened his body armor and helmet, and picked up his .45. The whole process took him less than thirty seconds.

Berks was still grinning. "I was just at the CP before I came to get you," he said. "Heard the grunts found a pickup with a 500-pound bomb rigged up in the back."

"500 pounds?"

"Crazy, isn't it? Lights out if one of those fuckers connects."

"Mm-hmm."

"Glad we got the night off?"

"Yeah."

"Gonna read all night?"

"After I lift. Also need to stop by the internet tent."

Ben went for a run earlier that morning when the platoon returned from its most recent raid. He took a shower, ate breakfast, and came back to his cot to read, but ended up falling asleep. When he woke up it was too late to call his dad. He felt awful. I'll email him tonight, he thought. Don't want the old man to worry.

*

About a month later, on a Saturday morning, Ed woke up at six. He made coffee and ate a banana, then sat in the living room and tried to read the paper. After an hour he gave up. I have to start getting some sleep, he thought. He finished his coffee, packed a lunch, and changed. At 7:30, he locked the backdoor and started walking towards his car.

The Saturday morning Mass lasted a little over thirty minutes. Afterwards, Ed shook hands with Father Kevin but did not stop to talk. He was too eager to hike in the woods. In the car, he turned on the radio, rolled down the windows, and whistled the entire way to the park.

Ed did a lot of thinking on the trails. He thought about the times he had gone hiking with Ben. He thought about work and the number of days Ben had left overseas. "Can I really take another two months of this?" he said out loud. He shook his head and started walking a little faster.

Around one in the afternoon, Ed sat down on a log and took out his lunch. He ate a few pretzels then studied the sandwich he had made earlier that morning. It was identical to the ones he used to make his son. Ed sighed. I really hope he reconsiders college, he thought. When he finished eating, he stood, dusted off his pants, and set off down the trail.

After hiking, Ed sat in the car with his eyes closed for about ten minutes. When he turned the key in the ignition, the clock on the dash said 6:17. He parked on Washington Drive about thirty minutes later.

It took Ed longer than usual to walk to the backdoor. His legs were sore from the hike and he wanted to watch the sunset. When he arrived at the house, he stood at the back door, rubbing his neck. Then he sighed, slid the key into the lock, and felt his hand start to tremble.

Inside, Ed performed his ritual. He walked through the kitchen and living room and stood at the closed curtains with his heart racing. Three torturous minutes passed. When he was ready, he grabbed the curtains with a shaky hand, closed his eyes, took a breath, and pulled. He looked.

"Oh, Jesus," he said.

Ed tightened his grip on the curtains and stared at the figure in the yard. He closed his eyes and turned his head. Then he looked again.

"Ben?" he said.

Ed ran to the front door and fumbled with the lock. His chest was heaving. He flicked on the outside light and stumbled back to the front window. Ben was still there, standing in the grass, smiling.

"I don't believe it!" Ed said. He waved at his son and hurried back to the front door. This time he managed to open it. He paused to look for a pair of shoes.

"Oh, who cares?" he said and swung the door open. In his socks he ran straight into the grass, his arms outheld, ready to embrace his son.

Only no one was there. Ed slowed to a walk and dropped his arms. He stood in the yard and looked around.

"Ben?" he said. "Benjamin?"

Ed walked to the side of the house.

"Ben? Are you there?"

Then he walked behind the house and around the other side to the front yard. He turned in place.

"Ben, please. Where are you?"

Ed stared at the front door. It was still open. He walked back into the house.

"Ben? Are you in the house? Please don't do this to me."

No response. Ed sighed and shut the door. With tears in his eyes he walked to the front window and looked one last time. His lips started quivering.

"Where is he?"

Ed backed away from the window and collapsed on the couch. He sat there with his eyes shut but only for a minute. A car drove past. Its headlights flashed across the living room. Pulling at his hair, Ed stood up, walked to the front window, and closed the curtains. Then he sat back down and wept.

Z.S. Diamanti

Among the Evergreens

The steam from old Mel's camp cup wafted and swirled in the light breeze. Day after day he'd sat there in front of his small wooden cabin. His chair was worn by weather, but he'd built it strong and it stood sturdy still.

He coughed, deep resonating coughs that shook his insides, and once he settled, he watched the forest and listened. He listened for flitting choruses of birds. He listened for the slow crunch of long elk strides. He listened to pinecones making their descent, crashing through branch after branch before reaching their final resting place on the ground below. He watched the way the trees danced around each other in the wind. He watched squirrels and chipmunks dip and dart in woodland games of tag. He watched memories.

There was a time when he walked among the evergreens, but he was too old for such things now. He used to stroll through the wood, quietly contemplating. He had always been a quiet man. But now he sat, holding his tin mug carefully and sipping thick black coffee, ever yielding its heat to the cool mountain air. Mel couldn't remember exactly when he had switched to decaf, but it was certainly after he'd met her.

After he came home from Vietnam, he signed on with a logging company in northern Montana. The Army had taught him two things: how to kill and how to work hard. Logging required the latter. Each day he and the rest of the choker-setters crew would go out after the fallers into the freshly downed groves and get to work. Securing each choker was difficult, and after Mel would finish one he'd take just a moment to breathe. He'd feel the grit and taste the sweat on his face. No other smell compared to the aroma released by those dozens of trees surrounding them. When the trees were all cleared, others from the company would come in behind them and plant new ones for future generations. It was hard work, but he loved it.

One of the loggers, Harold, was a homegrown local and he regularly begged Mel to come out to the bar with them. After months of pleading, Mel finally agreed to go. The bar was unremarkable, but he remembered everything about Margaret.

Harold had also convinced his sister, Margaret, to come to the bar that night, saying that she needed to get out of the house and get her nose out of all those books. Even through the smoke and neon cascade of light, Mel could see her beauty. She pulled her long red hair, straight as straw, out of her eyes and gently placed it behind her ear, where it stayed for only a moment. She never looked up from the bar. She was reading an old book, so worn that the paper cover barely held to the binding glue.

Harold introduced them with a goofy grin; he must have planned it. The two sat next to each other for a few minutes trying to take in all that was bombarding their senses in the small bar. Loud roadhouse music blared from the jukebox. Cracks rang from stick to cue to eight on the felt tables. Clinks came from the ensemble of bottles and glasses all around. Jokes and howls reverberated throughout the place. The smell of alcohol and smoke stung their nostrils.

"So, you work with my brother?" she yelled over the noise when Harold had finally stopped hovering.

"Yes, Ma'am. I do."

"So, you come here often, then? With the rest of these guys?"

"No, Ma'am. This is the first time. Harold has been . . . persistent, asking me to come here for months now."

"Oh? So, what made you come tonight?"

"Can't say really. I tend to keep to myself. I'm not terribly outgoing," Mel said with a slight grin and shrug.

"Me neither," Margaret shook her tattered book and laughed, the flippant strands of red hair falling back in front of her ear.

"I meant to ask what you've got there."

"It's a book. I thought loggers would know what books are," a wry smile scrawled across her face.

"Ha!" Mel laughed, nodding concededly. "I meant that I noticed it was really worn out. Having a few books of my own that look that way, I could only guess that it happens to be one of your very favorites."

"Excellent deduction, Mr. Holmes! But have you stopped to consider that it could, in fact, be a well-loved classic from the library? Surely, its popularity would see its continued borrowing from the library, and thus, its tattered exterior would be a result of many readers."

Mel looked at the book in her hand and then looked to her, trying to read her as he did his favorite stories.

"Perhaps, my dear Watson," he played along. "But perhaps it's a book you've personally borrowed numerous times." He looked down to the book again and saw his victory. "However, in the end my dear doctor, it appears that you failed to notice in your deduction that this particular book has no library markings at all! And therefore, we can conclude that my first deduction was accurate and that it is, indeed, one of your very favorite books."

Margaret pursed her lips and scrunched her nose trying to hide the smile underneath as she stared at Mel. When she lost that battle, she glanced down toward her book and thumbed through some of the pages.

"I could use some fresh air. Step outside with me?"

Without hesitation, Mel pushed his half-full beer glass away and helped her off the bar stool. As they grabbed their coats, and Mel helped Margaret into hers, he caught a glimpse of Harold with that same goofy grin watching them from one of the pool tables. Apparently, he was very pleased with himself.

That night they wandered among the evergreens talking of the stories that had taken them on exuberant adventures. They spoke of hopes. They spoke of dreams. She was even more beautiful laughing in the moonlight, her smile beaming from ear to ear. Snow fell all around, and when a rare moment of silence lingered between them they could hear the muted flakes touch the ground.

Their romance was fast and their love grew quickly. Soon Mel and Margaret married. He vowed to do everything he could to give her a good life. He bought some land in the mountains, blanketed with trees, and there he built the cabin—their cabin. And while they were building, Maggie became pregnant.

Stepping in the front door of the cabin with his empty tin mug, he glanced over to the old plaid couch in the middle of the living area. He could almost see her lying there with her large basketball belly, feet kicked up, head propped on the armrest, reading. She was so beautiful when she read.

She died the night John was born. Complications.

He raised the boy the best way a man knows how without the loving touch of a mother. They struggled in the early days, but John soon grew big enough to help around the cabin. The two began to function like a team, a dynamic duo.

Mel remembered the time little John asked about the book on the side table next to the old plaid couch. Maggie's old tattered version had finally given way, and instead of buying a new copy, she decided to borrow it from the library. Why did Papa always renew this book at the library? Why did he never read it? From the kitchen doorframe he could see it still sitting there now, *The Left Hand of Darkness* by Ursula Le Guin. He explained to his son that on one particularly difficult night after John had spent all day crying, and Mel could do nothing to soothe him until bedtime, he had opened it to the bookmarked page Maggie had been on when she died. He read only one line: "It is good to have an end to journey toward; but it is the journey that matters, in the end." He closed the book and even to this day he hadn't opened it again. He took very seriously his time with John after that.

One time, when John was just a boy, he found an eagle feather near the firewood stack.

"What's this Papa?" He asked.

"Hmm," Mel feigned. "Do we ever have a mystery on our hands. . ."

John's little nose scrunched the way his mother's used to and he looked perplexed at the giant feather.

And when Mel could hold it in no more, he announced in his best British accent, "And that, my dear son, is our first clue!"

John's big eyes grew wider in excitement. "A real mystery?!"

"Yes, my boy. The game is afoot!"

The two spent the whole day wandering around the woods finding different animal tracks, fur, feathers, and scat. The two wandered everywhere looking for the eagle that had dropped the feather, but never did end up finding him. Even though their trail went cold, the adventure had started something inside them, and they would have many more adventures as the years went by.

Mel and John did everything together. They read stories together. They grew together. They walked among the evergreens, breathing and feeling the presence of the giants around them. These giants were strong, and even after hundreds of years they stood sturdy still. The pair built each other up. It was out there that he taught his son to work hard. It was out there that he taught his son to live in and embrace the moment. It was out there that John became a man.

Washing his camp cup in the kitchen, Mel remembered the time John had spilled stain in the middle of the floor. Mel had taught him how to carve wood and make toys, but John was trying to impress a young lady at school. He'd been working feverishly on a wooden horse for her and accidentally kicked over the stain, spilling it all over the floor. When Mel got home from the mill, he found a desperate thirteen-year-old boy trying to scrub out the stain. It had seeped deep into the floor, and even now Mel could almost smell the lacquered memory of that day.

Eventually, as sons tend to do, John grew up and wanted to serve his country just like his father had done before him. A new war had started, a war on terrorism. After horrible events took place in New York and Washington D.C., the nation was reinvigorated and ready to defend itself once more. Some of the local boys were enlisting and John was going with them.

Mel was proud of his son, but he regretted not telling him what war was really like. Even still, John was a strong young man and he would have gone anyway.

As Mel left the kitchen, he looked at the old creaky door, the door that he almost always left open. He remembered gripping the doorframe, holding it for all the support he could get when a uniformed man came to his door to tell him that John had served his country bravely and performed his duty honorably before he was killed in action in Afghanistan. The frame held the weight of all his sorrow that day. He built this place strong and sturdy.

At the funeral Mel was presented with a posthumous medal and a flag for his son's acts of heroism. John had given his own life after saving a couple of fellow wounded soldiers and going back for more. It seemed small consolation to Mel at the time, but he truly was proud of his son.

Regrettably, the only time he got to see his son in uniform was in his casket.

After picking up the dusty old book from the side table, Mel exited the open door into the cool air once more. Another deep cough sent tremors to his core. He walked out on a familiar trail. No grass grew on it. It was well travelled. His steps were short and painful, but he walked onward still. He breathed in the aroma of the trees around him, one last gift from the gentle giants that had already given him so much in this life. He walked with his eyes closed. He did not need them. He knew the path.

He stopped in front of the two stones, one more mossed than the other. He wiped the dust, made sticky by the moist air, from the book and placed it on one of the stones. He reached into his pocket and pulled out a small medal, placing it on the other stone. A cold tear rolled down his cheek as he smiled at the two. He tapped both stones, said, "I love you," and made his way, step by step, back to the cabin.

Many years ago, Carol, the local librarian, had told him to keep the book, and he did. As he aged, it got harder for him to get out to the library, so Carol would send her grown son Lonnie out to the cabin with new books for Mel. He appreciated it. More so, he enjoyed his time with Lonnie. He was a kind young man and loved to share stories with Mel. He never dropped off books without coming in for coffee and a story.

"Lonnie, a young man such as yourself should be out finding a nice girl to marry instead of hanging around with an old man like me," Mel had once told him.

Lonnie turned away from the coffee he was making and toward the old wingback chair where Mel had been sitting.

"If I didn't come out here, who would make coffee and speak on the classics with you, old man?" Lonnie shot back with a smile.

"Hehe," Mel laughed. "I can make coffee myself, and when you're as old as me you talk to yourself, anyway."

Lonnie laughed as he brought two steaming cups of coffee out of the kitchen. He handed one to Mel carefully and sat nearby on the old plaid couch, cupping his coffee with both hands to keep it steady.

"I did meet a girl the other day, actually," he said.

"Oh?"

"Yeah, her name is Eva." Lonnie's eyes drifted as he said her name. "She's amazing."

The old man beamed, knowing well the feeling he read on the younger man's face. "Where did you meet this young lady, Eva?"

Lonnie went on to explain that he ran into Eva when he was bringing his mother some lunch at the library. Eva had been carrying a stack of books and Lonnie offered to help her. As he told the story, Lonnie may as well have been a space-travelling cowboy saving a beautiful Martian princess from the corrupted clutches of an evil warlord. Mel sat and listened. He did enjoy his time with Lonnie. The fervor with which he told his stories always made Mel feel just a little bit younger.

Lonnie would always recommend new books by new authors, and Mel would recommend old ones, many of which Lonnie had never heard. And the two would laugh about stories they'd both read.

The old cabin floor creaked as Mel pulled the chest out from under his bed and took out his uniform. After adjusting everything just right he sat at the old wooden table with three chairs to write one last thing in the journal he'd been keeping for Lonnie. He remembered the book, "It is good to have an end to journey toward. . . ." He had journeyed, and journeyed well. He had lived a thousand journeys in all his stories and one very quiet, but important, journey in life. Life had been hard work, but he loved it. He stood strong and sturdy all his days, but his time had come. Old age had finally caught up with him. He had been sick for a long while and knew this day would come. But he'd made peace with that. He also knew that Lonnie would be the one to find him. Mel would leave everything to him. He would miss their stories.

Before he laid down to rest he wrote one last instruction for Lonnie: "Lay me down among the evergreens."

Tanya Whitney

The Phone Call

I calculate the time difference and send her an email. "Are you up? Working?"

"Yes," she replies. "Give me about ten minutes to get to an open phone." A few minutes later the phone rings.

"How's it going?" I ask even when I'm sure I won't hear the real answer. After twenty years of marriage, I can tell when she's holding back information. When she's telling me what she thinks I want to hear. Not what's really going on.

"It's quiet tonight. All the aircraft are out flying, so I'm catching up on paperwork."

I can hear papers rustling as she speaks. "You didn't go out with them?" I ask innocently. She knows I hate when she goes with the aircraft. There's absolutely no need for her to fly. She's there as a maintenance supervisor not as a crew chief. She needs to stay behind the wire, though I would never say that out loud. Even as I know being behind the wire isn't a hundred percent safe either. Too many times our calls have ended abruptly due to the camp's coming under attack. Or emails have been shut down for days to control information going in and out of the base. Those are the worst, because you know something has happened but not to who or how bad.

"Nah, didn't feel like it tonight. They weren't going anywhere interesting or new," she answers wearily.

'Interesting or new' like she's on vacation, the thought crosses my mind. I know she doesn't take unnecessary chances, but I still worry. Things happen.

"What have you and the kids been up to? Anything interesting?" she asks with a hint of false cheerfulness.

"Nope. The kids are just going to school." I don't tell her about their school achievements or sports. She'll get upset because she's missing so much. Just give her the basics of grades and teacher reports.

"They're counting down the days until fall break already. You'd think they've been in school for most of the year rather than a couple of months. We'll probably head south for the week to check on the Moms." She doesn't know her Mom had a setback with her cancer treatments and has sworn me to secrecy.

"That sounds good. Tell them I'm still in Kuwait. Especially your Mom. She'll freak out if she knows I'm in Iraq."

We both chuckle over that because its true. My mother is a worry-wart while hers is a realist and knows she's in Iraq, though my mother-in-law pretends she's still in Kuwait when they talk. She knows her daughter too well.

"Listen, my minutes are almost up; let me say hello to the kids."

"Okay hold on. Robert, Rebekah, your Mom's on the phone!" Little feet scamper up the stairs, mouths arguing about who gets to go first.

"Stop right now. I'm putting you on speaker so you both get to go first." We go through this every time she calls home. I listen while they chatter back and forth for the remainder of the call.

When the beep comes signaling the call is about to end, I tell the kids to say goodbye and send them back downstairs. One last 'I love you' with all the cheer I can muster so she thinks, or wants to believe anyway, I'm not worried. But we both know I'm worried about her being over there. She's worried about being away from here. Neither of us telling the truth about what we are feeling or what's happening on our side of the world. The call clicks off. I hang up the phone and take a deep breath. Time to get back to our lives as we know it these days.

J.B. Stevens

A Wartime Breakfast with Ramirez

An explosion jarred me, and I opened my eyes and knew exactly where I was and there was no confusion. In the beginning of my wartime, when someone tried to shoot me or blow me up, it was disconcerting. After a while it became a part of life, and I almost forgot it was not a normal thing to be almost blown up, but I'd stopped caring. My feelings had stopped working.

I didn't like that they'd left, and I tried to fix it, but nothing had yet worked.

I jumped out of bed, grabbed my M4, put on my dollar-store flip flops, and ran to my grey-concrete bunker. Mine.

Ramirez was already there. He had a deck of cards. While everything blew up we played five-card-draw poker—no betting. It is less fun without gambling but much cheaper. I usually lose so this was a best-case scenario.

I pulled out my Skoal wintergreen chewing tobacco and filled the space between my lip and gums with the black shredded leaves. Ramirez gave me shit about the wintergreen again. I told him to stop, but he kept up. The minty smell mixed with the grey dust air. Airborne grit got in my nose and the back of my throat. It made my eyes water.

We talked about women, and Ramirez gave me hell about not having a woman back home. I didn't tell him about the tattoo-face girl I met on the last patrol outside the wire. Soldiers and locals didn't mix, and he wouldn't understand. But my existence was redundant, and I didn't care, and she made it less so.

The enemy fire kept oozing in. Soon, I heard our artillery's outgoing shots. After thirty minutes of back-and-forth-indirect-death-tag, it was over as if it never started and never happened.

When it was done, Ramirez grinned at me. "Breakfast?"

"Yeah," I said. "Let me get dressed."

We went to my housing unit. Ramirez held my rifle and waited outside. Inside, I put on my uniform. My boots fit tighter than normal, and I wondered why. I exited and we went to the chow hall.

The building had been hit by incoming, and there was a hole in the north-side corner roof and the north wall. I was glad the hole was far from the hot-chow line, that way no dirt got in my food.

I grabbed a brown plastic tray and got in line. Quickly, I got to the front. A Bangladeshi food-service contractor gave me scrambled eggs, bacon, pork sausage, a bagel with cream cheese, and pancakes. The food looked great, and I was excited and thankful.

I got a cup of black coffee and a glass of orange juice. The food reminded me of Waffle House back home—I'd gone there often with my father and loved the place. I sat down in a blue plastic chair that reminded me of grade school. Ramirez sat across from me; he had just bacon and eggs.

I pointed at his plate. "What's up?"

Ramirez slapped his not-huge-but-noticeable belly. "Going keto."

"Yeah? Why?"

"I got to look good when I get home." He laughed. "Keep the wife from cheating on me."

I laughed. "More, cheating on you more."

Ramirez stopped rubbing his belly. "What do you mean?"

"You know she's getting pounded by some fireman right now."

Ramirez frowned and forced a chuckle. The two Airforce sergeants sitting next to us laughed too loud and guilt bubbled in my gut.

"I'm just playing," I said. "I'm sure she can't wait for you to get home and isn't running around or nothing."

Ramirez winked. "Of course, no worries." He looked down and ate his eggs.

My guilt metastasized.

I held up my full coffee cup. "I'm going to get more, can I get you something?"

"No, I'm good," Ramirez said. "Thanks."

His voice seemed more alive and that made me glad. I didn't want to hurt my friends. I didn't want to hurt anyone.

I walked to the shiny stainless-steel coffee percolator. It was always perfect looking, and I wondered if some Bangladeshi guy spent every night cleaning the thing. A guy moves across the world to work for the mighty American war-effort and spends his days wiping fingerprints off a hunk of metal that does nothing other than pour hot-water over crushed beans.

I was a touch jealous, because that seemed like a good and simple and honest existence—maintain a thing that provides happiness to others. Simple and clean and righteous and no moral grey.

Everything I did, every day . . . did it make anyone happy? Did my bullshit make anyone's life better? Maybe being the coffee guy is more important than whatever the hell I was.

Since my coffee cup was full, the quest being a ruse, I poured the dark liquid in a trash can. I refilled my cup. The scent was strong and good. I loved that smell in the morning. I took a sip, and it was too hot but tasted great. The food and services were always top shelf at the base, and I appreciated it. I walked to the line and found a Bangladeshi food-service worker.

I made eye contact with the man. "Thank you," I said as I raised my coffee cup.

The worker, who was short and dark haired, with grey skin, waved and smiled. He was missing a front tooth, and the other was light green. His grin appeared genuine, and I wondered if he was the coffee percolator operator. I wanted to dig into the mystery, but Ramirez hooted so I walked back to the table.

Ramirez's eggs and bacon were gone, as was my bagel.

"I thought you were eating keto," I asked.

Ramirez shrugged. "Every diet needs a cheat day."

"How long you been on the diet?"

"This is my first day," Ramirez said. "Figure I get the cheating out of the way now and it is all smooth sailing."

"Smart."

"If you're gonna be dumb, you gotta be tough."

"I've heard."

"And I'm not tough."

"So you're smart?"

Ramirez tapped his temple with his right index finger. "They're playing checkers and I'm playing chess."

I laughed.

Ramirez pointed left. "That lady came looking for you."

I looked. Bassinette Reeves stood in the corner by the wall-hole. Light from the roof-hole shot down on her like a laser through the dust. She sipped coffee, and I thought of the phantom percolator operator. I stood up and walked over.

She shook my hand and held up her coffee. "You had any of this?"

I felt warm inside, another percolator appreciator. "Yeah."

She turned and threw the cup in the trash. "That's shitty coffee."

My throat got tight. "Sure." Wind rushed through the mortar-hole in the wall, and I hoped she would leave soon. "What can I do for you?"

She gestured to an empty table. We walked over and sat.

"I wanted to let you know," Reeves said. "Our investigation into Sergeant Golds's death is complete."

"And?"

She sighed. "It was a freak accident that occurred because he was raising drugged-up fighting scorpions."

"Seriously?"

"Yeah. We even brought in an arachnologist from the Smithsonian."

"What's that?"

"The Smithsonian?"

I frowned. "I'm not an idiot; what's an arachnologist?"

"A person with a Ph.D. in bugs. In this case, a specialist in scorpions."

"You brought an insect professor to a warzone?"

"We brought the doctor to Germany—we sent the body to an Air Force base there."

"I guess that makes more sense?"

"The arachnologist did a bunch of tests along with a coroner."

"And?"

"Golds was stung hundreds of times, it overwhelmed his system, and he died." A tear crept out of the corner of Reeves's eye.

"It had nothing to do with the scorpion fights or gambling or money or any of that?"

Reeves looked at him cross-eyed. "You think you're in a John le Carré novel?"

"No," I said. "Why?"

"There's not always some grand design. It's a fucking war."

"So?"

"So people die for stupid reasons, and it is done." More tears came. Reeves stood. "It's a fucking war."

I looked at the lines cutting down her face. "Did you know Golds?"

"I did. We both got here a year ago. I knew him . . . well." Reeves turned and walked away. She left through the chow-hall's door—even though the explosion hole was right there and leaving through an explosion hole is a once in a lifetime opportunity.

I thought about the tattoo-face girl from the last mission outside the wire and wondered if she was okay, and everything was bit better after thinking of her.

I walked back to the percolator, got another cup, and returned to my table. Ramirez had eaten my pancake and left, so my breakfast was over.

I exited the building through the bomb-hole and went back to my housing. I thought of Reeves's tears and the girl with the blue tattoo lines on her face.

I felt something.

Robert Morgan Fisher

You Stay Here

The sound was like God scraping a gigantic wooden match across the sky.

Shane immediately looked heavenward. The townspeople around him either ignored the sound or looked up with dull expressions of mild curiosity, shielding eyes with cupped hands. Shane noticed that in the time it took for Father to dash across the street to their station wagon, everything about him transformed. Dad still wore the same civilian clothes he favored on Saturdays: chinos, grey Naval Academy sweatshirt, worn-out sneakers, and Ray-Ban Aviators. But he'd changed. He became the mysterious professional Shane had only speculated about: the Naval Aviator who vanished every year, for months on end, to catapult jets off an aircraft carrier into parts unknown. This was a rare sighting, a thrilling thing to see: military-professional-Dad sprinting out of casual-Saturday-morning-Dad like a molting raptor.

Get-in-Get-in-Get-in! Father yelled.

Shane went from standing across the street to riding shotgun in one smooth motion. The station wagon peeled out in reverse before Shane had even closed the door. Three-on-the-tree transmission launched them into drive. They followed the downward trajectory of two descending parachutes to the outskirts of town; Shane and his father leaned out side windows like a pair of dogs. Shane scanned the horizon for a plume of smoke, telltale signs of impact.

Shouldn't we follow the jet? Isn't it gonna crash!? Shane squeaked.

Negative—it's an F-9!

Shane didn't know what F-9 meant, his knowledge of Naval aircraft being greater than the average citizen's but still somewhat limited. Father closed in on the parachute touching down in a bare farm field a half-mile distant, veering the station wagon off-road, bucking over furrows, sending up a rooster tail of dust. As they approached—triangulation of ejection, sighting, and recovery nearly complete—the parachute gently tugged a stiff body across the ground. Shane had never felt more exhilarated in all of his seven years.

Is he dead?!

Out of the car, running, Shane's heart thunked like a washing machine with an unbalanced load. Ripping off Ray-Bans, Father knelt over the motionless figure attached to the chute. Shane hung back, waiting in terrible anticipation, as if his father were about to deliver last rites. Shane prepared himself for tears as he watched Father's shoulders shudder. It took him a second to realize it was not a shudder of grief but a chuckle of relief.

Shane rushed forward and saw a mannequin in a flight suit, face painted gold. The effect was spooky and holy, like a golden idol had globbed off the sun and floated down to earth. Shane didn't quite understand. He pitched forward into the dirt clods, joined his father in a kneeling position.

Air show out at the base, said Father. Someone hit eject too soon.

Shane suddenly remembered: the weekend air show. They were supposed to go tomorrow.

Why is it a. . . Shane searched for the exact word. Mannequin?

It's just for show. He was supposed to eject in front of the crowd—remember last year? You don't punch a real man out of a jet unless you have to, he said. Dangerous.

Father shot up to full stature. On went the Ray-Bans.

You stay here.

He watched his father double-time to the car. Shane's heart was a fist knocking on a window, trying to stop someone from leaving.

Where you going!?

That other 'chute is the seat. We gotta recover it. He sternly repeated: You stay here.

Shane watched him drive to where the other parachute had landed, over by the high school. Now he was in charge, determined to do a good job. It was important, this assignment.

He was now Dad. The mannequin's gold face glowed. Shane reached out and touched the face and recoiled at the cheap roughness of its complexion. It was like touching the crucified Christ statue at church.

Voices and footsteps, people approaching. Unlike Shane's father, they parked politely along the road and trotted over to investigate. Shane solemnly guarded the mannequin. A fat man and his skinny wife arrived, huffing and sweating. Other people followed including a kid on a mini-bike and three high school boys. The high schoolers fondled the parachute, then thrust hands into the pockets of their jeans.

He's okay? panted the fat man.

The man's wife pointed. It's a dummy, dear.

He's not a dummy, said Shane. It's a mannequin.

The crowd clotted around Shane.

My father went to get the seat, said Shane, pointing. It was an F-9.

Huh? said the teenager with splotchy pimples.

From the air show. Someone hit eject by accident. Shane tried to sound blasé, as if this sort of drama were routine for him. He looked away with feigned annoyance.

The fat man nudged the mannequin with his foot. Shane fought the urge to slap away the shoe.

Your dad left you out here?

He had to get the seat, Shane repeated.

The what? said the skinny wife.

The seat. From the jet. It parachuted—Shane pointed—over there.

Oh, said the fat man, mopping his face with a handkerchief. Need any help?

Nope, said Shane. A red-tailed hawk rode invisible thermals above the bare field. Shane touched the mannequin's hand that, like the face, was also golden. He wondered if there was anything cool in the zippered pockets of the mannequin's flight suit, but didn't want to check with these people watching.

The high school boys smirked and kicked dirt clods. The mini-bike driver was about twelve and kept gunning the engine, trying to pop wheelies. He didn't ask or say a thing, just surveyed the scene and drew his own silent conclusions before tearing off through a windbreak of eucalyptus. The fat man and his skinny wife wandered back to their car, a little disappointed.

The three teenagers sauntered forward, attitude different now.

We'll take over from here, Bucky, the freckled ringleader said, spitting through a gap in his front teeth.

Shane said nothing. His insides went cold, legs numb and paralyzed. One of the teenagers had a wisp of a hair over his upper lip, a first attempt at a mustache that looked more like a misplaced eyebrow. He started gathering up the parachute and announced: Gonna cover my ceiling and walls with this crap.

Stop it! said Shane.

Whoa, Bucky, said the tall one. Who put you in charge?

My dad.

Where is he?—

He looked around.

—I don't see him.

Shane was too scared to answer. The tall boy knelt, breath rotten and burnt.

Come on, Bucky. Give up the dummy.

No.

Don't cry, okay? We ain't gonna hurt ya. But Ray here, he wants that parachute and I want the dummy.

No.

Ray had the parachute completely balled up. He dropped it near the dummy's head.

Come on, man. Let's split before someone comes.

Bucky here don't wanna part ways.

Let's see him try to stop us, said the pimply one.

The tall one said: Come on, Bucky. I want the dummy. Got plans for it. And o' course Ray here needs ten thousand pairs of silk panties.

Ray aimlessly threw a dirt clod and said, Shut up. I told ya. Gonna cover the walls and ceiling. In my bedroom.

Right. Your bedroom, said the tall one. Then to Shane, So, whaddya say?

Nuh-uh.

Come on, man, let's just take it, said the pimply one.

Give it here, said Ray, licking the hair above his lip and moving forward.

No! Shane felt his own face go hot as if he might cry.

Give you a dollar for the dummy.

It's a *mannequin.*

Ohhhhh! said Ray.

The pimply one mocked: A *mannequin!*

It's a dummy, kid. Dummy ain't even got any balls. The tall one kicked the mannequin hard between the legs and Shane felt a cold squirt of fear in his gut.

Stop it!

Gonna make me?

They started to pull the dummy away from Shane, but he hung on. Ray grabbed Shane by the hair and yanked. Shane's scream was drowned out by what sounded like a jet making a low-level pass. They all stopped and looked up at the empty sky. Then they saw that the sound was Dad's station wagon leaving the road and tearing across the field, bucking across remnant rows, crunching clods.

Run for it! said the pimply one. They bolted for their truck, laughing, and were speeding away by the time Shane's father arrived.

Who was that? said Father.

Shane couldn't answer, his words liquefied into a sob.

What the hell happened? said Father, looking over Shane's shoulder at the mannequin and parachute. Shane?

They . . . tried to . . . to take it. Shane hiccupped.

Stop crying. Who?

Teenagers.

Stop crying!

Shane wiped his eyes and nose on his sleeve.

You okay?

Yeah.

I got the seat. He opened the tailgate. Landed next to some tennis courts and these idiots didn't even see it come down.

Shane looked to the highway. Teenagers' truck was long gone.

Help me with this thing, said Father.

They loaded the mannequin in back, next to the ejection seat.

Can we keep him?

No, we can't.

Before Father closed the tailgate, he put a hand on Shane's shoulder.

Wanna ride in back? You can sit in the seat.

Yeah.

Get in.

Shane climbed in back. He sank into a silken sea—it and the mannequin floating atop the pair of foamy-white parachutes. His father closed the tailgate and jumped in behind the wheel. Shane held the mannequin's hand until doing so felt silly.

At the base gate, Shane's father bypassed the line of cars waiting to get in for the air show. The guard stopped them, even though the station wagon had all the proper decals.

Sir, I'm going to have to ask you to go to the end of the line.

Call the Ops Officer. You guys had a premature punch-out over town. You aware of that? He jerked his thumb, indicating the back of the car. I'm bringing it back.

The guard peered in at Shane sitting in the ejection seat next to the mannequin.

Proceed.

The guard saluted reluctantly.

Shane's father returned the salute and nodded, driving through the gate. Overhead, the Blue Angels executed a vertical climb, splitting away from each other like startled birds. The station wagon drove past Dad's familiar squadron hangar then veered left toward the control tower. They stopped before a run-down building with assorted aircraft out front. Shane's father parked, got out, and intoned, once again: You stay here.

Shane watched Father bound up the stairs. After several minutes, he reappeared alongside a young sailor who looked wary and confused.

The sailor's last name, CAMPO, was stenciled on his blue work shirt. They approached the station wagon and Father opened the tailgate.

Get out, Shane.

After his father and Campo had removed the mannequin, the seat and both parachutes, Father asked Campo what he was going to do now.

Campo shrugged: I don't know, sir.

Didn't anyone call about this?

No. Don't think so.

Where's the Ops Officer?

I don't know, said Campo.

Shane's dad frowned.

Do you know if anyone even went out to search for it?

Campo looked down at the mannequin.

It's just a dummy, he said. Sir.

I know that, sailor. But it landed *in town*. You mean no one even said anything?

The Blue Angels did another low-level flyby. Campo waited for the noise to pass, then shrugged.

Shane's father muttered: Goddammit. He frowned at the mannequin, started to say something more—then changed his mind.

Go back upstairs and finish watching the ballgame, he said. Jesus Christ.

Okay, Campo said, a little too agreeably.

They got back in the car. Shane waited for his father to start the engine. They sat there for a couple of minutes. The mannequin now rested in the ejection seat against the wall, atop the cloud of parachutes, like a crude caricature of an ejection.

Father said nothing during the drive back. As they approached town, Shane cleared his throat.

What's an F-9?

It's an old jet.

How old?

Dad didn't answer. Shane sighed.

Now what?

What? said Shane.

Why are you weeping?

I'm not *weeping*. And it was true. Shane wasn't weeping. But he was sad.

What's your problem?

We shouldn't have left that guy.

At first, he thought Shane was referring to Campo. Then he realized his son meant the mannequin.

We didn't leave him. We brought him back.

They could have kept the mannequin, dummy—whatever. Held him prisoner. Apparently, nobody had even missed the goddamn thing. And the seat too—those babies were expensive. How could anyone not be searching for it? No doubt Shane would've loved to have kept the seat, used it as . . . God knows what. But the dummy? Forget it. Frankly, it creeped him out. Like a corpse or something. Shane would've kept it in the garage—or worse yet, his room. Try explaining that to the squadron CO and his wife next time they came over for dinner.

No, the goddamn *dummy* was exactly where it belonged, and he'd just as soon never again set eyes on that jaundiced face.

Veda Boyd Jones

Change Orders

December 1, 1969

Coming to the Sig Ep House may have been a mistake.

I could have listened to the radio at my apartment, but the House had a TV, and I figured we were all in this together. Still, I didn't huddle with my fraternity brothers.

This was not the normal group, raucous before a mandatory monthly meeting was called to order. Tonight these college boys spoke in quiet voices as if they were waiting for a funeral. They crowded the two couches and upholstered chairs, which they'd rearranged to face one direction. They sat on the hard chairs that usually circled the card tables. They squatted on the floor and leaned against the walls, all focused on the flickering screen of the huge twenty-two-inch color television we had purchased that summer before Greek Week.

I leaned on the doorframe between the dining room and the lounge. The sharp wood cut into my back, and I shifted my position.

"Get it over with," someone yelled at the TV.

"Shut the hell up! Let us hear!" somebody shouted.

One guy scooted over and turned up the volume.

". . . in a bill signed just last week by President Nixon," the television reporter explained, "this lottery will determine the sequence of draftees for males born between 1944 and 1950. Nineteen-year-olds will be called first and should be inducted by February 1970."

Everyone in the room fit into that time span. I took a deep breath, trying to center myself while the luck of the draw decided my future. One guy fidgeted with a pencil, another rubbed his forearms with his hands, still another rapidly tapped his foot. Hell, none of my frat brothers wanted to go to Vietnam. Some were in college just for the student deferment.

On the drive to the House, I had wondered what it would feel like to have a low number or a high number. And I decided no matter which one, I'd act the same. Calm. Controlled. I wouldn't moan and I wouldn't cheer.

The TV reporter explained that men with birthdays corresponding to the first 100 to 150 draft numbers would certainly be called up. The next hundred were in the maybe category, and the final 100 would not be drafted and could get on with planning their lives.

"I'm counting on my birthday being three-sixty-five," I whispered to Tim Penning, who now hugged the other side of my doorway. He was the only other Sig Ep in architecture, so I saw a lot of him.

"You finish your project?" he asked. "I thought you were headed over to draw."

"Not even close. As soon as my number's called, I'll go." I smoothed my mustache, a nervous habit, and quickly stuffed my left hand in the pocket of my jeans. "But I'm counting on being here awhile."

"Me, too. Lowry over there?"

"Yeah, but he's got a medical. He doesn't give a damn about this. My deferment will be up in the summer."

"Shhhhh," came from several brothers.

On the screen, an old man stepped up to the giant glass jar, stuck his gnarled hand in and drew out a blue capsule. He handed it to another old man who took the capsule apart, removed the rolled up paper inside, and called out, "September fourteenth."

Another man took the yellow paper and posted it on a huge numbered chart. "September fourteen. Zero zero one." A male born on that date would be the first called by his draft board.

No one said a word as the next nine birthdates were called.

"December sixth."

"Shit!"

"You'll be all right, Cohill," someone said to the sophomore. "You've got two more years to wait it out."

"August thirty-first."

"Damn! Damn! Damn!" Tierney, a senior, beat his fist harder on the arm of the chair with each exclamation, which was extraordinary since I'd never heard him cuss before, while the rest of us put sailors to shame. He was a church-going guy, an anomaly here, but he said again, "Damn! Come on, Cohill. We're hitting Dixon Street. I'm buying." Tierney drinking? That was a new one. He jumped up and zigzagged through the bodies strewn on the floor until he reached Cohill, who sat on a hard chair in agony, his palms against his forehead, his head bowed nearly to his lap.

I watched as Tierney pulled Cohill to his feet, but still listened closely as more dates were called.

"June twenty-sixth."

"Wait for me," someone called in a cracked voice. I didn't even look around to see who it was. I concentrated on the TV screen.

The numbers kept coming. The voice kept calling out birthdates, which an old man in a suit posted beside the corresponding draft number on the big board.

"November twenty-seventh."

"Shit, that's close," Penning whispered. "I'm November twenty-sixth."

The numbers kept coming, faster now, ever faster.

"February twenty-third."

I gasped and my hand flew to my chest as if I'd just taken a bullet to the heart. I let out my breath in a huff and squeezed my eyes shut, then snapped them open to watch the hoary old man stick February 23 on the chart and say, "February twenty-third is zero five seven." Fifty-seven. Fifty-fucking-seven.

My life is over, I thought. *My life is over.*

So much for my stay-calm plan. I hadn't counted on the adrenalin rush that had my insides quaking. I smoothed my mustache. I took a deep breath to slow my hammering heart. Then another and another. My fingers worked on their own and smoothed my mustache again. The announcer had moved to other numbers while I gathered myself.

"I've fifty-seven," I said in a composed voice, over my internal shivering. "Well, I'm gonna go draw."

Tim Penning slapped me on the shoulder. "You'll be all right." He settled back against the doorframe, his gaze on the TV all the while he spoke. "I'll be there soon."

I shrugged into my coat and headed to the back parking lot, walking as normally as I could instead of running into the chilly night air. I'd parked the old black and white convertible on the hill for an easy getaway to the drafting room.

Fifty-seven. Fifty-fucking-seven. My mind repeated the number. *Fifty-fucking-seven.* I leaned against the car, lit a cigarette, and inhaled deeply. Drafted. Vietnam. How could I get out of this?

The old Volkswagen bug coughed a few times before starting up. It had been doing that a lot lately. My first year at the University of Arkansas I'd been afoot, depending on others for a ride. My sophomore year I returned in style. The Black-and-White was used, but it ran well. Through the semesters, I didn't cared for it like I should. I left the top down one night, and it rained. The next morning I drilled holes in the floor to drain the water, and we took it out and won second place in the intra-frat road rally. We got lost once on some Arkansas back road, but we made up the time with the next location, the Holiday Inn, and picked up a pack of matches, the required proof that we'd been there.

We would enter the Black-and-White again in the spring, my final road rally.

And then what? Then what? 1-A, that's what.

As soon as I graduated, I'd be drafted. Vietnam. Fifty-fucking-seven!

I parked across the street from Vol-Walker, the grand old library now the architectural building, and made my way into the enormous, dimly lit entry hall. On the left side, lights blazed through the two tiny windows in the heavy wooden doors that led into the drafting room. Sometimes if I couldn't sleep, I'd come over and put in a few hours on the board until three or four in the morning. It wasn't because I put off work until the last minute, although there was some of that, there was just too goddamned much of it.

The moment I opened the door, seven heads turned toward me.

"Hey, Bajollie, what's the word?" Lowry called.

"Fifty-seven."

"Damn!"

"Yeah. Damn."

"What about Penning?"

"Still waiting. He'll be over later." I turned on the light over my drawing table, second one in my row, and tossed my coat on the back of my drafting chair. I surveyed my drawing. What did it matter if I ever finished it or not? What if I was shipped to Vietnam as soon as I finished school in the summer? If I was killed, all these hours that added to days and weeks bending over this damn board were for nothing.

I picked up my mechanical pencil, studied my drawing of window details for the cultural center in Heber Springs, and drew a line so dark it snapped the thick lead.

"Shit!" I punched the end button to release more lead and automatically stuck it in the squat circular sharpener, grinding until I had to push the button for more lead. I threw the drafting pencil on the table.

Grabbing a wayward piece of flimsy, beige tracing paper, I crumpled it up and liked the crisp sound of destruction. It wasn't like breaking glass, but it was all I had. I tossed the wad into the barrel-sized trashcan at the end of my row, then although I knew it was wasteful, I tore off a piece of pristine tracing paper from the foot-wide roll, wadded it up into a tight ball, and threw it at the barrel. It glanced off the side into the aisle between the rows of drawing tables.

I ripped off another piece and crumpled it, then another and another. Three-fourths floated into the barrel, the others landing on the floor. Fucking terrible for a southpaw who'd pitched a shut-out in high school. God, that seemed like a lifetime ago, and yet Vietnam was on the nightly news then. I just rarely watched it.

A crumbled paper ball, this one bound by a rubber band for added weight, now sailed toward the trash barrel from Lowry's row. It missed the barrel by a foot.

"Ball game?" came Rod Dyer's voice from the doorway. He was an angular guy, tall and skinny, black hair and a heavy shadow beard, which could be full-grown in a matter of days if he didn't shave.

"Will Bajollie's fifty-seven," Lowry said.

"No shit? Want to move to Oklahoma?"

Dyer's dad was a high muckity-muck in the National Guard, and Dyer joined up two years ago when there was a vacancy. No way he was going to Vietnam, he'd told me. Dyer was a year older than the other fifth-years, and he was actually a fourth-year because he'd transferred into the program from some Oklahoma school. Something about a drug charge that went away, but Dyer said it was best to make a new start.

"My plan is to wait it out," I said, surprising myself with a plan. Could the war be over in a few months? For the first time since my number had been called, I had hope. A thin hope, but it was there, reaching out to me.

"Well, your plan right now is to get drunk. Come on." Dyer lifted his heavy eyebrows, which was something he routinely did to emphasize his words.

"I'm waiting on Penning."

"He'll know where to find us."

"Ten more minutes," I said. "He'll be here. They were calling those numbers pretty fast once they got started."

Dyer glanced at the wall clock. "Ten minutes, no longer." He moseyed over to his drawing table.

I sat in my drafting chair and wondered why I wasn't already at Neal and Lil's getting drunk. I didn't owe any great loyalty to Penning, even though we had roomed together for two years in the House before we'd both moved out to different apartments.

I sure as hell couldn't concentrate on window details. I tore a piece of gridded drafting paper from a tablet. It wasn't fancy stationery like Bar used when she wrote me one of her infrequent letters, but she said she looked forward to the variety of my writing paper, which was whatever I had handy.

I pulled out my drawer and grabbed one of the stamped envelopes I kept there, since this was the only place I ever wrote letters. In the even block all cap letters, the style required on blueprints, I printed her name, Barbara Heath, and her address at school, two states away.

Dear Bar,

I'm fifty-fucking-seven, but you know that if you watched the lottery tonight. I came over to draw, but as soon as Penning gets here, we're going down to Dixon Street and get drunk.

I knew she'd know my number. She may not have gone to the dorm lounge to watch it, but she'd listen to the radio. Last year she'd told me about the night a friend on her wing got the news her brother was missing in action. The friend cried for hours, and Bar said her outgoing personality transformed to hand-wringing anxiety. In support, Bar wore a silver MIA bracelet with his name on it and the date he went missing.

I don't want to go to Vietnam, and I will do anything I can to get out of it. I'm going to talk to Ed Beale when I go home for Christmas break. He made it there and back, and he knows all the ins and outs.

My neighbor Ed Beale had been drafted not long after he'd graduated high school. He didn't have a student deferment, and his folks didn't have the society standing to entice the draft board to skip his name.

Roger Pritchard had gone over about the same time, but he came home in a box. No! No, I couldn't let my mind keep going down that road.

I'll see you over break. When will you be home? We get out the Friday before Christmas. I'll call you and we can do something.

Love, Will

I'd find out when she'd be home even if she didn't write. Her sister, Karen, had married my brother. The family grapevine never failed me. But she would write about something as important as my draft number. We had a special connection. We were true friends, kissing friends.

"Time's up," Dyer announced.

"I'm ready. He must have a high number to be this long," I licked the rancid glue on the envelope and stuffed it in my pocket.

"I'm coming," Lowry said and grabbed his coat.

Once in the marble entry hall, I made a detour to the back wall to drop the letter in the brass mail slot, then our trio walked the block south to Dixon Street and turned east, heading down the hill. First stop was Neal and Lil's, a small dive, but one that I particularly liked. Neal and I became friends years ago after he showed me how to write out a check all the way to the end so no one could alter it.

The bar was full, every table taken by college kids, the males outnumbering the females by ten to one. The place had the usual smell of cigarette smoke and spilled beer.

"Number?" Neal asked from behind the bar. I took the only empty stool with the others hanging over my shoulder to order.

"Fifty-seven," I said.

"Have one on me," Neal said. "Anyone under a hundred gets free beer tonight. The usual Bud?" I nodded, and Neal looked at Lowry, who had just slipped his inhaler back in his pocket.

"Medical," Lowry said with a guilty catch in his voice. "Bud here, too."

"Hey, I'm already serving. Oklahoma. National. Guard," Dyer said, emphasizing each word.

"Lucky you," Neal said. He'd never taken to Dyer's swaggering ways. He filled three frosty mugs, set them on the bar, and collected money from the others.

"Thanks," I said. Free beer was nothing to sneeze at since I adhered to a strict budget. I wrote checks off my parents' checking account, so they could see every penny I spent, and Mom questioned every purchase. Now that Neal let me write checks for over the amount of my tab, I had some folding money in my pocket, but not much.

The bar buzzed quietly with talk of the draft. Numbers floated in the air. Boys slapped others on the back. The talk was subdued until more boys strolled in with celebration on their minds. I was on my second beer when Penning stuck his head in the door, spotted our group at the bar, and strolled in.

"What took you so long?" I already knew the answer from looking at him. His mouth was a straight line in sympathy with me, but his eyes were smiling.

"Three-oh-nine."

"You lucky sonofabitch," I said. For a moment, a dark side of me disliked him because he wasn't feeling my pain.

"Already called my mom and thanked her. Born six minutes before midnight. Six minutes later and I would have been number forty-seven."

"You lucky sonofabitch," I repeated right before someone opened the door, letting in the sound of POP-POP.

"That's a gun!" someone shouted, and college boys swarmed toward the door.

I stayed on my barstool. It didn't sound like the echoing growl of a shotgun, which I knew from weekends hunting quail with my dad, my brother, and my grandpa. Besides, if somebody was out there with a gun, I didn't need to be out there, too.

Someone opened the door again and shouted, "Tierney shot his toe off! Stupid ass."

"What the hell?" Dyer rushed to the door. Lowry was right behind him. Others followed.

Neal wiped the countertop as he moved down the bar toward me.

"A kid shot his toe off?"

"Fraternity brother of mine. Brain of a retarded tick. His number's something like eleven or twelve."

"Don't *you* do something stupid."

"I'm not into pain." I drained the last of my beer. "I guess I'll go check on him." Only three other guys remained inside, and one laid his head on the table, totally wasted. I did not want to get into a discussion with Neal about the war. Talking to the Korean War vet was like talking to Dad about the flag and country. I was *not* in the mood. "See ya later. Thanks for the beer."

I stepped outside and watched two police cars, lights flashing, sirens blasting, racing down the hill from the highway, followed a block behind by an ambulance. Students, mostly males, crowded the street, and cars that could go nowhere in the madness jammed up. Someone held up a scribbled sign on poster board. STOP THE STUPID WAR. The real crowd milled across the street on the parking lot of George's Majestic Lounge, Dyer's favorite drinking hole and the place where holier-than-thou Tierney had shot the gun.

The crowd parted as the first police car pulled to a screech. Under the streetlight, I glimpsed Tierney, grimacing with anguish and regret and shame, sitting on the curb.

I knew then, with absolute certainty, I would not avoid the draft.

Photography

Photography Winner

Christian Tillett

Combat Boots

Photography Honorable Mention

Randy Brown

Prepping for "The Box"

Photography Honorable Mention

Ruth Snipes

The Well

Christian Tillett

Patrol

Christian Tillett

Camels

Niko Eden

Men on Mars

Joseph S. Pete

USS *Indianapolis* Survivor

Joseph S. Pete

On Deck of the USS *Indianapolis*

Joseph S. Pete

USS *Indianapolis* Commissioning

Randy Brown

M-RAP Rear View in Kapisa Province

Randy Brown

Just Over the Wall (at National Training Center)

Poetry

Poetry Winner

Harry Gordon

The Wounded Cry Out for Their Mothers First

Driving away in the old Chevy
--after she'd walked into the little corner market
on 2nd and Lime with my little brother in tow
to ask for a job—my mother notices him chewing gum.

"Where'd you get that?" knowing he doesn't have a penny,
not since our stepfather disappeared.
As if he'd cast his eyes on Medusa,
he turns to stone while I dissolve in the back seat.

A U-turn and we're parked again in front of the market.
She hands him some change.
"Go back in there, tell them what you did,
and pay for it."

She plucks him from the car, nudges him through the door,
and stands there arms crossed in her best print dress
until he comes out wailing in that tight striped shirt
wiping his nose on his sleeve, his cowlick flared over his forehead
like a feather in his cap.

I don't know if this is my brother's worst day
as he still has Vietnam ahead of him,
but I have a feeling he's been hit
by something he hadn't seen coming,
like the invisible rounds between tracers.

As they walk back to the car together,
he reaches up for our mother's hand
without looking,
knowing it will be there.

Poetry Honorable Mention

Jessica M Granger

WE FORGOT

the defensive, dangerously low tide of Inchon Harbor that won MacArthur fame.
the combat pilots flying so closely soldiers could feel the heat from misdirected explosives.
the sole first black soldier to report to his newly integrated white platoon.
the dragon's dream of metal mortar shells buried in shallow graves beneath the Pusan Perimeter.
the nervous breakdown of front-line soldiers who were told to "Stand or Die."
the Southern Baptist girl who became a liquor-thirsty nurse without water.
the collapse of the Hun River bridge destroyed by friendly fire while innocents fled across.
the shiny, squeaky cerulean skin of limbs that snapped off in the cold, left inside socks on soil.
the dead men stacked in truck beds whose bodies were stacked in pyramids that twitched.
the trenches they died in as they were filled in by new growth of a brighter day in Spring.
the sailors buried at sea, sent afloat in cedar boxes to dance with the sea kelp of temperate water.
the peace treaty that was never signed by a neutral hand without stakes in the land of strategy.
the Superpowers that broke a nation at the 38th parallel shortly after the Second World War.
the documentaries that recited acceptance of a body count if the number follows the word only—
the way the face cringes at the headline "Only 20 dead today" on every printed newspaper.
this was never a war.

Poetry Honorable Mention

Ryan Stovall

Army Brat

summer's last faint breath pushes up the hill
a gentle warmth that speaks of growth
and love

there will be no other breeze
like this
until some time late
next year

the tall grass bends
embraces the breeze
the oak's leaves give up a gracious sighing
the girl's short yellow hair ruffles
a little

black clouds
are gathering behind the breeze
stark brutal winter
staring out of each

the girl turns and with a solemn heart
trudges down the meadow hill
that faces out
across the sea

Ryan Stovall

The Firing Line

under the searing sun of Gao
between some scrubby brush
and the last few stunted desert trees
at the edge of the sand
where the paved road and all life ends
a sweating line of coal-black sons
prostrate themselves upon the ground
as if to pray

all squint
against the searing sun of Gao
across a swath of sand
toward a jumbled mound of basalt
lying broken black and pink
at the base of a long cliff
jutting skyward like a church

this Mecca they all face
is the rusting hulks of cars
not Humvees or APCs
but sedans and minivans
strewn haphazardly before the rocks
and baked by the searing sun of Gao

at a order the hammers fall in unison
the reports echo from the cliff
to join the cacophony
of whining ricochets and plinking tinny hits
that roll sharp and menacing
across the desert lying bare
under the searing sun of Gao

atop the cliff
the ghost of Mother France
looks down on them
from the bombed and burned out remnants
of an old colonial villa

tears flowing down her peachy cheeks
to wither and parch
beneath the searing sun of Gao

Victoria Elizabeth Ruwi

Blue Shrapnel

Blue beneath your skin,
pieces, not in peace,
small scalene triangles,
war angles raising up
years afterward. My fingers
trace arcs around shrapnel,
my touch avoiding stretched
skin jacketing fragments.

Bits of non-equal sides, 180 degrees
locked in, pucker your skin before
they can be safely removed, detached.
Calmly, we do not speak of the bomb,
we un-share our battles, we crumple.

Discharged, you strangle camouflage.
I, newly promoted, wrap in it.

Victoria Elizabeth Ruwi

35 cents for Steve

Cans clunked from break room machines
as we gulped down caffeine and fantasy
with jests and the kind of cursing meant
only for soldiers of shared experience.
Spit polished boots, stiff uniforms,
rank centered on hats resting on tables,
we shared our dimes and quarters,
escaped from that civilian world
of 50 cent soda, were rewarded
for protecting our country each time
we bought Coca-Cola for 35 cents.
We all had our reasons for losing
ourselves in Army pride and games,
but those kinds of stories were saved
for the twentieth hour of twenty-four
hour guard duty, when the crowd slept
and two soldiers had no more jokes.

The drill was no different when Steve
and I guarded the barracks, watching
our duties and time dwindle down.
All rifles mounted, all locks secured,
drunk soldiers evacuated to their rooms,
we stared into the pentadecagon dark,
listening to pings break the unsounded night.
Steve squared his hat, sipped his soda,
finally mustered his voice and his story:
He heard a tiny click, then that thunder;
he feared thunder when he was five.
His mother had pointed the same gun
at him just before she changed her mind.
They said he was covered in her blood,
but Steve could not remember the blood.
He could remember when he was seven
how the red flames burst from his father's
little plane, how it exploded after take-off;

he remembered the firm hug his father
gave him before leaving on that plane.

Stillness merged the ending of those hours.
In the twenty-third hour we unlocked doors,
woke-up troops, gathered the quelling light.
In the twenty-fourth hour, as cans began
dropping, we awaited release and sleep.
For Steve that meant a cot in the barracks.
For me it meant home, where my husband
sneaked in morning kisses, nested our baby
in my arms, then left to begin his own duty.
After he died, silence filled rooms I entered,
with everyone buffering themselves
from me, a twenty-five-year-old widow,
no longer a welcomed party guest.
Except for Steve, who brought me
35 cent sodas from the machine,
read silently at my dining room table,
and waited for me to choose to talk.

Victoria Elizabeth Ruwi

Next Exit

Refusing the hundred bucks to do
what he said his wife was too
ill to do, left me two miles west
of where I began, still in San Diego.
Next ride, a doctor bought me dinner,
tried nothing, warned me about other men.
Off I-8 a trucker tried, shrugged off
my no, let me drive his eighteen-wheeler
on the long flat stretch across Arizona.
Farther on, a different trucker heard my no,
drove the shoulder at 55, shoved me out,
bounced my backpack to the dirt with me.
I leaned on the *Welcome to Tennessee*
sign as a stuttering sun buzzed my road
burnt pebble pocked marshmallow skin.
That night I slept in a bathroom stall,
head bobbing in the air, butt on the toilet,
feet up on the latched door, just to be alone.

So I joined the Army, where I was never alone.
Drill Sergeant Huss made me talk to a wall
because I laughed when she yelled, made me
talk to a tree for saying *Hi* to some guy,
put me on KP for the hell of it. The Specialist
in charge of KP asked, but I peeled potatoes,
mopped floors, cleaned gunk from tables,
smirked when Ponce creeped off with him.
MacKay, Sego, and I scrubbed pan after pan,
listened to music, because it was forbidden,
snuck out early to watch a formation of men
raising their rifles, then grabbing their crotches,
yelling: "Here is my weapon. Here is my gun."
We searched for guns until their drill sergeant
eyed us, counted no stripes on our shirts, shouted
"You privates drop!" Before we went down, I
asked, "But Drill Sergeant, where are our guns?"
He made us kiss dirt till darkness knew dark.

Kissing dirt became habit as we grew muscles,
became soldiers, received orders to move on.
MacKay and I arrived at Fort Gordon,
gave blood together, just to get the day off.
Giddy from an oxygen withdrawal fog
I told her how my parents moved me
away from my first love in Virginia,
how tempting slow dances with Tom
led to hours of exploring, how he
and I became clever in our teasing.
Gazing down from the top bunk,
MacKay said she'd never been in love.
I may have glanced up when she added,
"until I met you," but I was drifting off.
Through the months our room filled
with women, barefoot in fatigue pants
and white T-shirts, most looking to escape
the *psst psst* of men trying to conquer us,
some women plying to blemish the innocence
in Mackay's porcelain face. When MacKay
shipped out to Germany, my orders sent me
west to Fort Riley where I received a letter
from Tom, postmarked at Fort Gordon,
saying he drove all night to find me
in Georgia. He missed me by two days.
Tom never came to Kansas.

Victoria Elizabeth Ruwi

Walk

Watching me, you said maybe I'd relearn
to walk like a lady now that I was out,
but a soldier once trained has no return.
Only in drill or surrender is an about
face waged. This left-right walk, no doubt
IDs me to sister warriors with dog tag lore:
who call cadence for every woman's clout,
who wish for ceasefire at their very core,
who with weapons train for war, grabble gore.
This breadth of female grace eternally fits in,
marches every boot step, the HUP-two-three-four
of battle, of peace, of pain, toes frostbitten.
Walk with me, without quarrel, pride in tow:
Neither think of me as goddess nor foe.

Joseph S. Pete

Cincinnatus

Cincinnatus beat back the invaders
and then his sword into a ploughshare,
or so the hoary sketch of a story goes.
So too did Sergeant Davies of Dayton
relinquish his M-4 to the soft-palmed pogue armorer
after he was freed from the sweaty grip of Iraq's swelter.

Sergeant Davies returned to college on the GI Bill, caroused,
married, bought a low-slung house on a dead-end street,
took up a hand fork and trowel after the workweek's toil had exhausted itself
and tilled his mangy yard and flinty garden,
trying to fill in the loose lattice of dirt patches
and impregnate the potter's soil with darling cucumbers and plump-cheeked tomatoes.

As he smoothed the ready mix of grass seed, mulch, and fertilizer
over the arid dirt-pocked lawn that had been long neglected,
he wondered if those checkpoints he ran before Iraq's first election,
if all the aggressive barking at befuddled motorists just trying to make their way wherever,
planted a seed that would germinate and flower
into some efflorescence of a legacy that would justify all the suffering in the suck.

As he thumbed tiny seeds so prone to skittering off into the loamy soil,
he thought of his battle buddy, the acne-ridden kid he used to play
Grand Theft Auto with back in the barracks, the friend
he followed around to all the bars, who was now packed
under a cold, unforgiving slab of still dirt,
after all the prerequisite pomp and ceremonial ho,nors.

Loretta Tobin

I Hear Birds

I want to hear a pair of seagulls
voice their klaxon warning wail,
see them bow to each other,

then strut and waggle and mew,
before they hop and run, like planes
on a runway, flap their wings—

take to the air. When I hear gulls,
I can inhale the tangy, briny salt
smell of the sea, watch the waves'

soft lap approach my feet,
threaten to douse my toes
with a chill cold drink.

All I hear are crows.
Big, black crows. They line up
on wires strung between power poles

and along the edges of buildings,
soldiers ready for inspection.
There's no mew, no wail—

only a shove or two, a dress right,
dress, and loud caw caw caws.
Here, even the crows know formation

drills. I scoop up a handful
of sand, hold it near my face,
sniff and hold my breath.

Where is the sea or the salt?
All I detect is the universe,
its *gotcha laugh*—

like a murmuration
of prayers
and desert heat.

Kristine Otero

Free(dom) Onion

Smells of PTSD and pride permeate
the Outback at lunch.
Free Bloomin' Onions
on Veterans Day.

Real Heroes—WWII and Nam
give my onion to those boys.

I should have been
an unauthorized Barbie watch
a tiny scorched boot.
Photographs and memories
remnants of erroneous greatness.

Yet here I stand
an untried war criminal—
the prosecution rests.

Driven by poverty
not patriotism.
In nightmares narrative shifts
amalgamating moral injury.

Sycophantic civilian salutes
for fraudulent freedom.
Echos of undulated honor.

Every November 11th
Bloomin' free(dom) onions
soak up the stench
at the Outback.

Jay Harden

Give Yourself Peace

These are my thoughts of deeds
Dark dressed in words:
A uniform does not make actions right,
Nor do shiny bars command righteousness.
My life was lost again and again
As it took others again and again,
Clouding all the while this heart.
One life taken is too many.
As duty becomes mechanical
Prosecuting war,
Care flies out the cockpit
And self-worth free falls below
Crashing and burning jungle.
No one like me comes home again,
Staying lost in that sucking swamp
Of doubtful mind.

Only I can redeem myself.
And her words come back to me,
Soft and clear,
Unchanging and unerasable:
You, the one I always love,
Are an innocent man,
Then, now, and forever.
Men akin to you
Are the heart and soul
Of this country.
You can give yourself
The only gift that matters.
Give yourself peace.

Reflections

Jay Harden

I reflect on my life
And I think
I bloomed too late.
I am too old to be vital.
All my cherishing was then.
I live alone, I am alone,
Within a loving extended family.
They are busy becoming
And I have become and gone.
I wonder about my death;
I am curious about the moment of it.
I hope the women I loved
Will be waiting for me somewhere.
But I worry they have grown
Beyond me,
Gone on to something better.
I wonder if oblivion awaits
In the next life,
If I will even know it is so.
A hopeful part thinks
That can't be all there is.
But will I know?
No one has reported back on the next life,
Perhaps because there isn't one,
Perhaps because it is not worth reporting,
Perhaps because it is too glorious
Living in that magnificent continuing instant.
One word described her: enthusiastic,
En theos, out of God.
One word describes me: curious.
She was a teacher of little ones.
She taught the eternal love of learning,
Teaching me, too.
I relish feeding my brain,
My love of life, my love of you.
Sure, I have some regrets,
But none of consequence.

My life is purposed,
A journey without destination,
A journey in light and shadow,
In peace and war.
I write, I paint, I photograph, I play music,
So the world will know this boy
Who lived in spite of circumstance,
Then the man who lived that way, too.
I leave behind evidence
For anyone who cares to notice.
I can do and be as before,
Still doing and being:
All this beyond battle,
Or because of it.

Jay Harden

Safe Sleep

War murders sleep,
The sought sleep
Of kings safe in their realms.
Safety and good sleep
Go together like blood brothers.
No matter how far away
Or long ago your war,
Sleepless memories appear
Like a surgical scar
Until they stir you
Toward today.
Coming home
Brings safe sleep,
The unknown peace
That wakes you satisfied.

Aubree Tillett

Coming Home Soon

Holding the handle to military luggage
mailed home: hope
that you would be holding
my hand soon. Opening the black trunk,
wafts of burnt hand-rolled tobacco,
sand grains, and stale air escaped,
after polluting unwashed camouflage uniforms,
used Christmas gifts, preserved
birthday cards on top
of stacked souvenirs bought on base.

I unroll one of your worn beige shirts, slipping it on
over my head, wondering about these new scents,
wondering what you've experienced when I know
you're sworn to secrecy. When we talk, we are strangers
speaking different languages, trying to be polite. Trying
to be honest without open conversations,
we connect via short texts and crafted care
packages, monthly mailed. It is a relief receiving
this piece of you home to hold.

This evidence proves
this year was just temporary
and soon a memory.

Wes Smith

The Box

Creased cardboard rough and worn
from passing man to man, each
contributing an item: sunglasses,
a photograph of wife and kids or maybe
mom and dad, others leaving
keys or a condom, like teenage boys
holding the collection plate at church,
a letter they wanted to read
one last time, as though those words
would carry them into combat.
When I hold the box, I give up
what I hate to leave behind but can't lose:
the coin my grandfather gave me
when I shipped out, telling me
he carried it through World War II:
a 1920 silver dollar, its surface
rubbed bare, a talisman
he hoped would protect me
as I fly 30,000 feet above
Iraq, where the plan is to leap,
land safely, kill people, and get out
with my ass still attached
because no one's more important
than these fourteen assholes
I eat, sleep, and bathe with every day
and who I will cry over
and who will cry over me
if we don't make it back to that God-
forsaken patch of concrete and sand
in Georgia where we'll drink
and fight, pass out before being
reassigned again in order to
keep us from going crazy and losing
what's left of us so we don't
end up inside another kind of box.

Wes Smith

Nocturne

I'm awake—the bed shakes
as I bleed out, alone, a blade still
buried in my thigh.

I feel the warm wet
on my legs but it isn't blood.
I throw the sheets in the washer,

retreat to my favorite chair.
Flipping through reruns,
I settle on a comedy I've seen.

It's dark. I hear his breath
wheezing slow. The odor
of cigarettes as he drives the blade

deeper. I scream—my dog barks.
The windows blush:
I'm on the floor, the TV

flickering the news of a new day.

Wes Smith

Camo

The first small dog and Dachshund I've ever loved

Few understand how close
we are; you've saved my life
often, sometimes daily.

Your gigantic personality surpasses
your true size, wrinkly skinned
stubby legs, and slim wagging tail.

You see my zoned out look
or my anxious, shaky demeanor
and you're there, distracting me.

You look at me, asking permission
and then, when I say no, giving me
a pitiful look until I give in.

When the dead parade across my mind
and I don't know where I am, you bark
and head-butt me, knocking them out.

The bark that bellows from tiny lungs
startles people and makes larger dogs
stop and take notice of you.

You stay with me as my breathing
slows, I focus on you, scratch
your ears and let you know I'm back.

Everyone is your friend, you greet humans
and animals alike with a sniff, receive
a scratch, a pat, or a sniff in return.

Many times, I rush home, pick you up, hug
you, my tears wetting your coat, your chin
resting on my shoulder makes my day better.

You prance when you're excited to go
somewhere, dance on hind legs for cheese
and make everyone feel loved.

Heikki Huotari

Peach

The serenade is suitable. On hearing car alarms I care. Your calling sunrise peach won't disallow my doing so. The fountain pen of age is leaking into my shirt pocket. Spontaneity, is that what you'll be wearing to the sacrifice? The serenade is suitable. On hearing car alarms I care. No Loch Ness Monster gets in my way. Armchair executioner, what inadvertent innocence will you be weaponizing now? If thy glass eye delight thee pop it in. A broken biorhythm is correct once every other day. The serenade is suitable. On hearing car alarms I care.

Lucid

Heikki Huotari

If they were ashes they were snowflakes and they might have melted on some tongue so what about the difference between vertical and horizontal don't they understand? As in the trail of bread crumbs there's a fork, a joke, two stories of an origin, if poor or poor but honest twins are switched with twins at birth event horizons dilate like an iris and subtended angles of invisibility like opportunities abound. To see if you're asleep try turning on and off the light, to see if you're a hero too.

Heikki Huotari

Portmanteau

To the naked man and woman God said, Who said you have one iota of redeeming social value? Bounded by tautology and contradiction, fun was had by all but twenty. If you crossed a picket line which was invisible then the ekphrasis that you sang was seething. Islands of potential energy, the gods are bumping fists. Five sheets of cardboard on the concrete and you still can't sleep? Don't pay attention to that pervert, Rorschach—symmetry is where you find it. Keep your eye on that expected value, Esperanza—sum the products of the outcomes and their probabilities. You're just the messenger and when the rapture happens you'll be busy training your replacement.

Randy Brown

wearing the tree suit

After liberating a dog-tag machine,
Cadet Hensley was a new man.

He punched the keys and stressed the metal:
Last name, first name, middle initial.

Religious preference? *Druid.*

During field exercises thereafter,
whenever the chaplain came around,
he would immediately run into the nearest woodline

to hug himself
the largest tree he could find.

Carl "Papa" Palmer

His Limbo Soliloquy

Actually I like lockdown. I already was before COVID anyway,
but now I've got my privacy. No family feeling forced to visit
or hold vigil in my netherworld, he confides through the phone.

Both of us former Army soldiers placing us on common ground
made introductions easier with the usual "where were we when"
comparisons of duty assignments all military members embrace.

Though sharing multiple telephone calls these past seven months
since my assignment to be his companion as a hospice volunteer,
I have yet to meet him face-to-face due to pandemic restrictions.

Using his bedside number at the nursing home I can call anytime,
not worry about visiting hours, ask if he's busy, got time to talk.

His answer's most always the same, *Just busy here being alone,*
too close to death to complain. Clicking me to speaker he begins
what he calls "me-memories from a time when when was when."

Mostly musing of being anywhere but there, lost in an actual place,
blurring "what was with what is" behind and in front of his shadow,
recalling dreams as a younger man, of a future in past perfect tense.

And times talking of present times from his no man's land outpost,
All days end as they begin in purgatory, today recopying yesterday,
cared for by hosts of faceless masked angels not letting me die alone.

Forgive me only thinking of myself, I just need you to hear I'm here.
Inside I'm your age, the two of us sharing a brew at the NCO club,
years ago and oceans away, comrades-in-arms talking of our day.

To me he's the sergeant with permanent change of station orders
in transition for his final mission ending his time on active service
in hopes his God is religious and his terminal assignment is good.

Valerie Elizabeth Young

Suffer in Silence

So many people dealing with the fight within
Mental illness on the rise, suicidal ideology: don't know where to begin

Isolation from the world feeling so alone even in a room full of people
Society needs to understand I'm not a glass so I'm hard to see through

When my thoughts run rampant or I'm feeling at my worst
Military, life stressors, it's like my weapon went from automatic to burst

Variables assisting my thoughts to a plan of suicide
Alcohol, drugs, peer pressure, stress in everyday life I want to run and hide

Decoy presented: an the outside things APPEAR to be fine
However, internally my mental, my sanity, my health all seem to decline

A smile is the expression of the face NOT the heart
Built up emotions, Feeling all alone, Isolated, wondering am I doing my part

Suicide: a temporary problem to a permanent solution
Contemplating is there a resolution

Suffering is worse than dying the pain is relived and surreal
Not even the normal veteran can deal

It's important to reach out to a support system: there we will find balance
Remember YOU are NOT alone so you don't have to suffer in silence

Pat O'Regan

The Child

In Memory of Lt. William Little III

She was four when he graduated from West Point.
She remembers the day—bright and happy—
He in his sparkling uniform,
She feeling pretty and the focus of attention, too—
The two of them posed for a photo:
"I'm really something," she thought,
"Just like Billy!"
How could she ever forget the day?
Or forget the time he took her to a parade
And she stood on the seat of his Corvette
Looking at the floats through the roof,
Or at the house on the Jersey shore,
When they all swam in the ocean,
Daddy, Mommy, and Billy,
Teaching his little water rat to dog-paddle,
But she wouldn't remember when he,
Nurturing of heart, came in late,
And found her awake in her crib at bedtime—
"I put her down to sleep!"
"She's not tired yet, wait till she's sleepy."
How she giggled in his attention!

Love flowed freely through the family—
Billy and his dad were especially close—
Parents and two siblings—all were predestined
For a lifetime of loving and happiness together.

Then he was gone, off to war,
Writing letters to him was all she had,
Sitting with Mommy at the kitchen table:
"I hope you are ok."
Reading his letter just for her:
"A special thought for my favorite girl
From the jungles of Vietnam. Happy birthday!"

But the matter at hand,
The martial imperative, commanded his attention,

For he loved soldiering, too,
Leading men in battle,
A perilous love.

The risk is the thing,
Only focus can reduce the risk,
Attention is demanded.
As he moved through the jungle with his men,
Only brief thoughts of her came to him,
On tip-toe, as it were,
She crept into his mind, just now and then,
Though a letter from home
Brought a brief burst of delight:
"My water rat is progressing!"—
Filling his heart with thoughts of her—
Before, back to the business at hand,
With an iron focus and determination
To accomplish his mission.

He would be a soldier to the hilt,
Even in the face of death—
For that's what soldiers do.
It was his passion and his calling,
As if from above, and he would
Answer that call—to the hilt.

Good officers stand face-to-face with death.
Brave beyond question, they must be—for the men.
In the face of hardship and danger,
Leading, inspiring, and compelling others to follow,
Even at their own risk of death.

What thought of his family, of his sister,
Could say, "Be careful, they want you home"?
What thought could find a place in his head,
To turn him aside or urge caution,
When he faced a bunker complex
Charged with enemy soldiers?
Other than his mission and his men,
What other thought could he have?

Before the martial imperative,
She could only wait for the next letter,
Or the rare phone call,
Or the homecoming, sometime,
Or never, though that could not find a place
In a mind so young.
In his place, the moment was all—
The bullets zipped past,
Cutting through the air around him,
With a sickening whine,
Or lashed the ground nearby, whip-like.
Taking charge, as men fell wounded,
He risked a fusillade of fire
To break the enemy's position.
What must I do here and now was all he had.

Of all careers, soldiering commands attention.
The child could only wait,
And find a little innocent luck—
Just a little luck. . .

He and the enemy were stuck,
Face-to-face with each other.
Blocking fire and gunships
Held them in place. His mission determined
That he stay. He wouldn't be the first to blink.
He would fight this battle out.
His bravery was challenged,
And fear was no match.
Any thought of leaving the field
Could find no place in his head.
Home and family were far from his thoughts.

War is all chance.
Attention, discipline, and training only help,
Some must die.
Move here—live. Move there—die.
Look up—live. Look down—die.
Or vice-versa.
Her luck ran out.

"Mommy, police or soldiers
Are coming up the walk," the child said.
"Go up to your room!"
"Mommy never sounded like that before.
What did I do?"
Coming downstairs at last—
"Where are they?"—
To find them in the kitchen,
Hugging and crying.
"Billy's not coming home.
He's never coming home."

What now?
Live even more, live like mad,
But remember there was a brother
Who loved and was loved,
And who made your heart sing,
Who lives on in your memories of him,
Who will always hold a special place in your heart,
As you live your life to the hilt—loving and happy—
Just as he would have wanted you to do.

Charity Winters

Pre-Deployment Training

The military always war games for contingencies found in plan books of fantasy equipped with material left over from wars before. The purpose is preparedness, to know the mission, tactics, and gear, but pre-deployment locations and deployment destinations rarely have anything in common other than the military way of hurry up and wait. They say the principles of land navigation are the same whether in the Rhineland or Kuwait. Still, I am not sure how the forests of western Germany prepared us for the Iraq insurgency or which war we were preparing for in Korea as the bogeyman north of the DMZ got testy with his long-range weaponry.

The best training was in Texas. The landscape was like Iraq. In both places the summer heat blistered lips and heads. Then there was the outhouse at the live fire range with a dead animal floating inside the sludge of human excrement that medics swore did not cause all the open sores reported during sick call, but still they cordoned off the shack of shit. Combat first aide training got real when one of us face-planted in an anthill. And, not to be confused with where we would deploy, there were people on the run, border-hopping, oil, snakes, bugs, and in Texas everybody had a gun.

Jonathan Tennis

Chatting with a Vet

Please don't
Thank me for my service.
I didn't do it for you.

Please don't
Ask if I was afraid.
Because I was.

Please don't
Say that I'm your hero.
I'm not the stuff of recruiting posters.

Please don't
Decorate my grave since
it's the only time I see you.

Jonathan Tennis

The Lord's Prayer

Our Father,
Who aren't in heaven,

Hallowed be thy big blue marble.
What's left after the flood?

Kingdoms come and go—
Few have ever conquered Afghanistan.

Give us something to eat
Other than MREs or foot bread.

I've forgiven those who trespassed.
Will my trespasses be forgiven?

I don't have time for temptations here—
Eat, sleep, work, gym, repeat.

As I wait for that plane to deliver us from this place.
This—The Graveyard of Empires.

Ben White

Memorial Thoughts

37 years.
Another anniversary
of a bad day at the 1st and 51st Infantry
when the M60 machine gun was accidently
discharged and fired a round out of the barracks
room and through the arm of a soldier
and into the collar bone of another, where
it was deflected down to tumble around
his chest cavity, taking out his heart and
lungs just like it was designed to do
                                        before lodging
                                        in a rib—
the medics swarmed on the Thursday
morning, 26 January, and kept his
brain alive, but when the German
doctors at the *Krankenhaus* cracked him
open, they showed them the mush
of organs they'd been trying to save. . . .

T.M. Hudenburg

familiar statues

to Sophocles

even breath falls short this time
a road sign what words you decipher
describe a long-ago battle fought here

only silence now save the wind
says it all with a whiff of cordite
gunpowder the breath of a long aftermath

the dead
and the wounded
strewn over there

opposing armies led by brothers
each courageous both killed in action
the one deemed traitor goes unburied the other receives a hero's burial

Antigone casts dust upon the former
what the law of gods demand
but what of the laws of man

Creon is certainly no god
who rules this land
the tapestry in his throne room unravels

squandered between sunlight and growing shade
furthers the unwinding
fear for Antigone, sister to both, caught in familial duty

controversy darkens again this time at Creon's command
Antigone too will die
as will her beloved Haemon, Creon's son

perhaps we wish to avoid such tragedy
when our time comes
could we not expect some deus ex machina

cast a bit of dust
save us a bit of rest
in peace until then we have our orders and move on

Milton P. Ehrlich

The War to End All Wars

I'm awakened early in the morning
by a blaring siren on an army truck
hauling a billboard with Uncle Sam
pointing his finger at me, screaming:
I WANT YOU FOR THE U.S. ARMY!
A loudspeaker reminds everyone
the draft has been reimposed, and
all able-bodied men are required to
report for duty at once, including
veterans regardless of their age.
Hey, wait a minute, Sergeant—I'm
90 years old and can no longer soldier
the way I used to!
Don't worry, Grandpa, we will assign
you to the heavy artillery drone division
where you can stay in bed with pajamas
on and demolish anyone who is not behaving.
The Pentagon's code name is: The US Army
of Genghis Kahn—the plan for a war to end all wars.

Donna Aycock Meares

His-Story, V-J Day, 1995

Ninety-five degrees.
The barest breeze plays one note on the wind chime.
A cicada in the pine anticipates the storm.
On the porch, I await the rain.
I smell it first, then see its spatter in the birdbath.
Across the leaden sky thunder marches. Its rumble
must remind my father of the great guns, for
he joins me on the porch with
a frayed, black book I've never seen.

A blade of light whitens his face.
A leaf quivers on the dogwood.
Large drops thud the roof.
My father opens his book,
reads his words penned half a century before.

I see him then,
thin, shorn, wearing khakis,
scared, stacked five deep
on a transport ship crossing the Pacific.
I see his comrades, Tobin, Stockard,
playing poker, smoking Luckys.
I feel his heart jump
on night patrol in Saipan's shadows.
He transits to Guam on way to Okinawa,
in gnawing wonder of what lies ahead. . . .

The rain falls straight down.
My father pours out memories. Young soldiers
bond in duty, swallow fear,
prepare to invade, to stare down death.
Lightning cracks close by.
The ultimate weapon
drops
Japan to its knees,
brings home my father,
who waits fifty years to share his story.

Robert Brewer

Ode to Admiral Zumwalt

Here come the good humored Squids,
thick pockets bulging and cash laden,
boot camp survivors in Chuck Comiskey's park
where Bobby Bonds puts one in the cheap seats.

Executives we groom,
minions managing the lower decks;
administrating the turnstiles,
halt they say to the hotdog vendors,
popcorn purveyors, distributors of selzer
and beer. What about my friends
who sit on their hands recession bound?

No matter you've never met them.
You meet them now.
Hand them out the sausages,
let them guzzle gallons,
and send in the Goose, gossaging a fastball
they can only glimmer.

Then clean up the paper wrappers.
Deposit fecundities in Dixie cups.
Fix your thoughts on challenges,
I pay you for everything.

Robert Brewer

Regret

The word has come.
I, you see, have a college degree.
You could be a chump. You could be a champ.
Everybody goes to boot camp.

Will you wait for me?
Can you suspend our dalliance?
Will you put in abeyance your pink brassiere,
the one with light-hearted spangles?
Could Ghandi quit his teaching,
Lenin his assassinations?

The sweat is cold, the moonlight dim,
in 40 years necrotic scabs have yet to heal.
Careers have ended now,
promotions granted and denied,
stairways ascended leading nowhere.

In my long running nightmare,
her hollow answer quivers

"Yes."

St. Leger "Monty" Joynes

Night Messenger

I know from waking
at three a.m. with a troubled mind
and lying abed in violent thoughts
until I have to get up
and drink good whiskey
to dull and quiet my ego,
and turn to the reading of scriptures
that the span of existence
is not to be compared
with the span of non-existence.

I become relative to myself.
The whiskey and the approach
to God makes me kin
to all poets of all time
and partner with all composers
of great music.

We weep to create a pathway
to truth. We sit alone
in the middle of the night,
realizing the limits of self
and the potential of the divine.
We are the messengers.
We are the hopeless runners
with ultimate pain on our faces
who must deliver or die.

St. Leger "Monty" Joynes

October 1965

We used to sit around the office,
an orderly room at Fort Polk,
and make signs to confuse the privates.
A personnel clerk created:
"Please don't feed the Arklatex."
But the most profound was:
"Remember, bullshit killed cock robin."

We did not see Wilfred Owen
waiting in a combat uniform
for the order to move just
a few hundred yards to death.

Nor have we seen men in flesh
or blood from unspeakable wounds.
Weapons we know, and some of the sounds,
minus pain, but not dying cries.

We are now on orders and wait,
hearing rumors of wounded men
who ate last month in our mess hall
and were drunk in town before they left.

We laughed too much toward the last
and planned trips with leaves
and passes we never got and
stayed mostly sober until the last night.

Jason Arment

Transformations of Aggression

By every door I keep a club
Transition can be hard
To stomach & hard to carry & hold
Like keeping my head up after a victory Pyrrhic

When I got back from Any extended time with the USMC
I needed a weapon, even if a BB gun or a knife, by the bed
For me to sleep. Now I don't need anything
To sleep, because what troubles my dreams
Isn't by the bed but in my head

I feel like moving quickly
As dancing through the cathedral of myself
All the stained glass turning kaleidoscope
As I spin across marble stone and so much gold

Of course there is a quartet playing softly
So the cries from the dungeons aren't sussed out
There are always the unknowable places
Mine are filled with actual skeletons

But you see, I never talk about it
People who had known me their entire
Adult lives were surprised at what they read
Because I'm not here to entertain you or
Sell you T-shirts that say "I HELPED
DESTROY IRAQ" followed by a bunch of Japanese
Characters that don't even make sense translated

Jason Arment

Born Again Hard

I've tried, & I'll try again
To find the fruit the Good Book spoke of
"Ye shall know them by their fruit"

Usury, the logical end-state of our system
Along with ever shrinking dowagers
Fetish mining politicians &
Bitcoin mining white supremicists
Drunk on the idea they have of themselves
Which is interesting, like how all war memorabilia
Is also murderabilia by default

One of the things the salty veterans don't tell you
Is how war memorabilia differs from murderabilia
In that if you don't keep the war memorabilia
People will convince you the war never even happened
Just like the guns used to talk back & forth on the line
Knives have their own walk & talk, people do as well
But only one of those will lie to you
Please don't count me among them

If our fruit is corrupt—rotten, flyblown, or worm bitten
We are hewn down & cast into fire
"Wherefore by their fruits ye shall know them"
& there will be times their destruction will be the will
Of the lord, & we will be his instrument
Wading into the fire as we burn

Ben Weakley

Heat and Pressure

I.

Awake to the slanted light
on a March afternoon
in Baghdad, I rise
from sweat dried into crust,
salt stained in the shape
of a body on a poncho liner.

A man died today.

Knelt down beside a garbage pile
to wipe his forehead,
to rest the weight
of armor and ammunition
on his bones before

the fire clap and dust cloud.
An echoed thump,
felt not heard,
inside a steel tomb.

The urgent voices,
radio squawk and chatter.

send doc quick he's fucked up bad

Punctured tire. Spalled armor.

patriot x-ray   patriot x-ray
this is sapper five
medevac request follows
litter   urgent   surgical

II.

Sergeant P says
you cannot place blame
when someone is killed
because war is about killing.

Sergeant O says it's best
not to imagine
trading places.

Sergeant J says he didn't hurt.

he didn't feel   anything
just heat and pressure
and   like that—

Snap.

everything went black

Essays

Essay Winner

Rob Silverman

Incident at Range 27

An addiction to Ancestry.com led me online to North Carolina, to Cumberland County. I did not find my missing relative. I saw Bobby's death certificate instead. Bobby was not a relative; I just knew him. I watched him die in Cumberland County fifty years earlier. The death certificate called it an accident. You decide.

Bobby arrived dead at Womack Army Hospital, Fort Bragg, from a gunshot wound in the chest. The army doctor who performed the autopsy wrote on the death certificate, "white male, twenty-two years old, unmarried." Under the cause of death, the doctor wrote "accidental discharge of M-14 rifle." The doctor did not report how the accidental discharge happened, but I remember.

Bobby died in August 1969. Five hundred Americans perished in Vietnam that month. The army barely noticed one training death in North Carolina.

Months earlier, Bobby might have imagined dying in a firefight or a mortar attack. Then he got a spot in the Massachusetts National Guard. He was not going to Vietnam. Bobby had only to finish training, spend six years at weekend drills and summer camps.

In six weeks of training, we had learned how to march, salute, spit-shine our black boots, and shine our brass belt buckles. We learned to get by on five hours' sleep, swing through an overhead ladder to eat in the mess hall and disappear on Sundays to avoid picking up cigarette butts.

We spent most days learning how to shoot someone a hundred yards away with a rifle and how to eviscerate someone up close with a bayonet.

Finally, we spent a week practicing everything we had learned near a replica Vietnamese village. The huts had thatched roofs, and live goats wandered the hamlet, little bells around their necks jingling. I sat on the dusty red earth against a large tree, breathing the piney air stirred by a light breeze, staring at a jellylike lump of canned chopped eggs with ham. A goat eyed me. I offered her the chopped eggs with ham. She turned her head away to eat a plastic pouch of cocoa powder.

We marched to Range 27, fell out on the double, took seats on bleachers. We faced pairs of trails separated by brush-covered berms running up a gentle hill. "Walk to the top of the slope one row at a time," a sergeant said. "As targets pop from the ground and the brush, drop to the prone position and shoot them. Walk at the same pace. Don't get in front of the firing line. Set the safety on your rifle before standing up."

M-14 assault rifles were heavy, clumsy. The safety was on the front of the trigger guard. If you did not look, you might squeeze the trigger instead of the safety.

Row one moved out, shooting at targets as they appeared. Cracks of scattered gunfire and the sour smell of spent ammunition filled the air. Each row behind followed.

Range sergeants carried safety paddles—the tops of large food storage cans stapled to flat wooden sticks painted white and red. If the sergeants held up the white side, it meant to move and fire; if they showed the red side, that meant stop moving and cease firing.

Near the firing line's center, I sat on the fourth row. A soldier on the right trail in front of me fell behind his row. Bobby was to his left.

A range sergeant supervising that pair of tracks smacked the back of the slow soldier's helmet with his safety paddle and yelled at him to move faster. The sergeant poked the slow soldier in the back with his safety paddle and ignored Bobby. Bobby moved ahead, dropped to the prone position.

A moment later, Bobby remained motionless on the ground as his row got up to start forward. The red clay turned dark under Bobby, turned dark from the blood spewing from his chest. I did not distinguish the shot that hit him. I doubt that Bobby heard it; they say you never do.

The senior range sergeant ran along the bleachers, waving the red side of his safety paddle in every direction. "Cease fire! Cease fucking fire."

The trainee on the right stood, froze. The range sergeant who paddled him took the trainee's M-14 and shoved him until he fell. Then he kicked him. Other sergeants grabbed the range sergeant, held him back.

I bobbed my head, rubbernecking as if passing a car crash. I could not see Bobby behind the sergeants standing around him. This is the army, I thought. We are at war. The military takes casualties in battle. Range 27 is as close as you get to combat before actual combat. Expect this.

A sergeant walked over to us in the bleachers, his shoulders slumping. "That stupid shit," pointing to the bewildered soldier on the right trail, "squeezed the trigger instead of the safety. The round went just above the ground. Hit the other guy in the chest. A million-to-one shot." I looked at the ground between the trails dotted with little mounds of rocky dirt. Any might have diverted the bullet. None did.

Bobby's friends cried. I did not weep, nor was I holding back tears. Nine years earlier, my father died from cancer at forty-two. I was twelve and cried for days. Then I stopped crying for the dead.

A sergeant with a prominent Adam's apple shouted at the ones who cried. "This is a lesson! You fuck up, somebody dies! You pussies! You'll see a lot more of this where you're going!" Another sergeant led the wound-up sergeant away. We stayed quiet, waiting for orders.

I recalled a story my grandfather told. During World War I, a young man who served in the US Army came back from Europe's trenches with medals and received a hero's welcome. His friends asked him to describe the war. His answer, "They give you a rifle. A guy could get killed!"

Three Huey helicopters landed in quick succession. Bobby left in the first one; helicopter number two took away the soldier who shot him.

Helicopter number three delivered a lieutenant colonel. He faced us, standing tall, shoulders back, hands clasped in the small of his back, feet apart. He explained what had just happened. "This was an accident, men. Military police officers are on their way and might ask what you saw. You saw an accident."

The military police never asked.

Sergeants covered the spot where Bobby died with fresh red clay and ran the rest of us through Range 27. Next to that spot, I dropped to the prone position, fired, hit the pop-up target. It fell backward, then sprung forward. I turned my head toward the fresh red clay for half a minute, a moment of silence for Bobby. A sergeant told me to keep moving. I set the safety on my M-14, stood, walked up the slope.

We bivouacked that evening in the woods a short distance from Range 27. Bobby's friends organized an impromptu memorial service on a knoll in the piney woods. Bobby was Catholic, but the service was ecumenical. One of his friends said to me, "What do Jews say when somebody dies?" Jews recite the Mourner's Kaddish, a long prayer I had not uttered since my father's death. I stumbled through it in Hebrew. Only the handful of Jews in the company understood what I was saying, but the rest liked it. They bowed their heads and said, "Amen."

Essay Honorable Mention

Anne Boaden

UXO

Circle fifteen hundred feet overhead

Peer through the eyepiece of your Night Targeting System into a chilly January sky. Watch Afghans prep cooking fires, ride motorbikes, tend to skinny goats in the dust. Normal pattern of life this evening. You feel like a voyeur, protected by distance and technology. That's your job as a front seat pilot in an AH-1W SuperCobra attack helicopter.

The radio crackles. "Hacksaw, Medevac in progress. Needs escort. Standby for grid." It's DASC, the controlling agency that supports immediate tasking.

This is new. Your stomach flutters.

"Hacksaw, copy," replies your aircraft commander over secure comms. He's the commanding officer of the west coast skid squadron currently deployed to Camp Leatherneck. He's been an easy signer to fly with today.

"Heads down."

Your helmet mic transmits your voice automatically over the intercommunications system. No pushing buttons. Handy.

You grab the Zebra pen dummy-corded to your kneeboard with a fraying piece of string. Should really replace that soon; loose objects in a cockpit invite mishaps.

Standing by to write.

DASC relays the coordinates. Your signer punches them into the Control Display Navigation Unit. Basically a boxy '80s black-and-green computer. Scribble down the numbers as backup. Write large enough so your chicken-scratch handwriting, compounded by the washing machine thumping of the bird, is legible. Don't be embarrassed by failing to be able to read these notes during debrief. Again.

"Heads up."

Two pairs of eyes are better than one.

You and Dash Two check off station and turn toward the new grid. Both Cobras accelerate to max chat. Things might get kinetic.

Look out the front canopy at the passing landscape. Four months into this tour and it all seems pretty much the same. Brown and dry. Then again, it's better than being stuck behind your desk at Wing headquarters. You appreciate combat flight time, hours that will be entered into your logbook in red pen. Combat equals red time, respect.

Orbit overhead ten minutes later. Two Blackhawk helicopters circle below. Angels of salvation, dark as shadows, ready. Heard nothing but good things about these Army guys, utmost professionals. Dustoff is the go-to Medevac asset in Helmand Province.

Switch frequencies.

"–cking land! Land now! Medevac! Help!" A Marine screams in near-panic for a chopper.

"Calm down. Landing zone needs to be cleared first," a Dustoff pilot replies. "Give us smoke for a mark."

Dustoff is laid back and assured. This guy's good.

"Yellow smoke marks LZ." The Marine's voice steadies over the radio.

Flip your visor up and return to your sensors.

"Heads down."

Scan the area around a squad of infantry Marines. They're bunched next to a litter with a man sprawled on it. He's Urgent: the Marine needs care within two hours to save life, limb, or eyesight. Speed is of the essence. This Medevac's personal—he's a fellow jarhead.

Wisps of yellow canister smoke waft into your line of sight. Zoom out to search for bad guys. They must see the helicopters by now.

"Hacksaw, stand by for 9-line."

Booya.

The Joint Terminal Attack Controller of the infantry unit has taken over comms.

Adrenaline engulfs your body.

JTACs are specifically trained to control aircraft in combat. Pilots work with them all the time back home. This is a no-shit, kid-gloves-off request for close air support. Your first live 9 line out here. Giving permission for both choppers to fire lethal rockets and guns, Hellfire missiles if necessary. No more notional run-ins, battle damage assessments, fake target destruction. This is actual combat.

It's game time.

The 20-millimeter cannon hangs directly underneath your seat. Two chubby Hellfires bristle from the right-hand wing stub. Seven high-explosive 2.75" rockets wait patiently in their pod on the left. Your job is to obliterate the bad guys. You've trained years for this.

Don't fuck it up.

JTAC transmits his 9-line. Read back mandatory lines: 4, 6, and restrictions. Feel like you've downed ten cups of coffee—focused, excited, jittery.

Today might be the day, your combat cherry popped. You're looking for the Taliban fighter who ambushed the grunt patrol. There's at least one casualty, the Marine, and possibly a local national child to boot. The Taliban doesn't care about collateral damage.

Smoke swirls from the middle of a dusty road below. Temporary LZ. Dash Two flies 180 out in the overhead so either of you can tip in and shoot baddies. You've practiced this formation hundreds of times in North Carolina. It's natural.

Nothing suspicious down there, though what counts as suspicious remains unclear. No visible weapons anyway.

Fixate on a Dustoff bird sinking into the LZ that's ringed by Marines. They're taking a knee too close to the landing spot. Make it bigger, boys.

You should be looking into the distance for trouble, beyond the hasty cordon. But you've never seen a real Medevac before. Marines carry the litter toward the bird as it lands in clouds of brown dust. Glance at the time. The second Army bird loiters nearby, out of the way.

A garbled transmission breaks the silence. You only make out "UXO."

Unexploded ordnance.

"Did you catch that?" you ask your aircraft commander.

His reply is over the radio, "Confirm UXO situation."

Silence.

Then, "Casualty has UXO in his leg. He's been hit by an RPG, but it didn't explode." The JTAC seems calm about this.

What?

An image flashes through your head. A metal rocket-propelled grenade protrudes from the Marine's leg, green and cylindrical. Bone fragments are scattered everywhere. Blood drips everywhere. Agonizing screams echo through the dirty air.

It's fucked-up. No shit he's Urgent. Can Dustoff even let him onboard?

No hesitation. Another mushroom of brown as Dustoff thunders straight-line to the closest Forward Operating Base, Edinburgh. Look at the clock: Dustoff was on deck for less than two minutes. The second Blackhawk continues lazy circles, waiting for the next patient.

Scan the ground for bad guys. Nothing. The Marines on deck don't move much. They're still guarding the hasty LZ. Now empty LZ. There's radio chatter about the local national casualty. An Afghan child, a girl.

Look past the Marines, farther up the dirt road. The sun is low in the sky. Dust seems thicker. It's hard to make out details. Some local nationals ride around on motorbikes, skirting the edge of the camera's vision. Hard to tell their intentions. The juddering helicopter doesn't help.

A motorbike turns down the road toward the Marines. Cursory drive-by for Taliban intelligence before dinner? Free pot-shot at stationary Marines? Curious innocent commuter? Zoom the sensor in to check. The bike swerves away. It's hard to see who has weapons from a thousand feet above.

"Bingo fuel," your aircraft commander transmits across the inter-flight frequency.

"Same," Dash Two relays.

You pull away from the sensors, ring marks indenting your eye sockets like sucker punches.

"Heads up."

The AFT FUEL LOW caution light is illuminated.

Quick glance at the fuel gauge. The needle bounces between hash marks. Hard to take a reading. Schwag it halfway. It's low regardless.

Normally, helicopters fill up in Bastion: swing by the hot pits, top up without shutting down. You can't make it that far. That's twenty minutes away. Need to follow the Blackhawk into FOB Edi and grab gas at the Forward Arming and Refueling Point. It's no big deal; even you've done FARP operations before.

Your section checks off-station and requests hot gas at Edi.

"Hacksaw, negative. We have UXO on site." The Edi controller is trying to follow the rules.

"Edi, Hacksaw low on fuel. UXO understood. Risk accepted. Request permission to land."

Your signer must be rolling his eyes in the rear seat. Of course there's UXO. The Medevac was an unexploded RPG buried in some poor Marine's leg. The section circled overhead while he was flown out. You saw the litter being loaded.

"Hacksaw, negative. Refuel at Camp Bastion. Edi is closed."

Turn inbound. Can't make Bastion. You need fuel or you'll fall out of the sky. Yet another casualty.

"Edi, Hacksaw critical fuel."

The FWD FUEL LOW caution light flickers on. Land as soon as possible.

Cross over square Hesco barriers filled with broken rocks and rubble that define the boundary of FOB Edi. An American outpost in the middle of nowhere. Inside the wire now. Committed.

As the skids settle into dirt, the controller grants you permission to land.

Both birds hot refuel with rotors turning. Nobody leaves either cockpit.

Look out the Plexiglas canopy and watch people scurrying around the Blackhawk a few hundred yards away. Wonder what's happening, if the Marine's onboard. Try to imagine the pain of a rocket stuck in your femur. You can't.

Hope he makes it.

The Cobra's air conditioning blasts ice crystals in your face, but sweat continues to roll down your back. Amped from the sortie and possibly finding the bad guy. Today might be the day you schwack an insurgent. One who tried to kill a Marine with an RPG. One who injured an innocent civilian child. One who's still out there.

You're thrilled and terrified.

Boom back to the LZ a few minutes later. The second Blackhawk continues his orbit. JTAC tells you there's an Afghan girl who's been shot by the Taliban. She needs medical attention. He says a man, maybe her father, is driving a motorbike with the girl on the back. The Marines don't know where he went.

Start searching.

"Heads down."

The motorbike from earlier, where is it? The circle of Marines guarding the LZ passes through the sensor's field of vision. No one's moved. The light is fading, sun sinking through the jagged horizon. The dust and haze of evening combine into a pilot's worst nightmare; you can't see shit. Switch from direct view optics to the forward-looking infrared. Try to pick up warm bodies and hot engines. Toggle between white-hot, black-hot, then white-hot again. A trick from your skid-kid buddy, Chris.

Movement.

Slew the sensor left and find a motorbike, a white blob with a blurry smudge on top. Toggle back to optics and zoom in. Definitely someone on a motorbike. Definitely heading toward the Marines.

"Motorbike. At least one person onboard, heading south," you say.

Your aircraft commander can see where you're looking through his head-up display in the rear seat. A green carrot shows him where your Night Targeting System is pointing.

"Contact."

Watch as the motorbike creeps down the road. Is the girl onboard? You can't tell. This light sucks. The motorbike stops a hundred yards from the Marines.

"This might be our guy," your signer says. You agree, but say nothing.

The driver sits on the stopped bike. He doesn't dismount. Switch to the infrared sensor again. This time see two distinct heat sources, one smaller than the other.

"There's a second body on the bike," you say. "Maybe a child."

A sudden cloud of dust encircles the motorbike as the driver whips a U-turn. Heads back up the road.

What the hell?

If this is the father with his injured daughter, why is he driving away? Surely the Marines have been speaking to him on the road. Telling him it's okay. Bring the girl. We'll save her. We have a helicopter waiting.

The man must have seen the first Medevac leave with the injured Marine. Must see the other Blackhawk circling nearby. He must know the Marines are here to help, whatever lies the Taliban tells locals.

"Heads up."

Radio the JTAC: "Pretty sure that's our bike with the injured local national."

Roger.

"Dustoff, low on fuel." The second Blackhawk is about to turn into a pumpkin. You look left and see the black bird flying donuts above brown fields. Waiting.

The sun has set with rays of gold and auburn streaking toward you through the dust. Pinky time.

"You have the controls," you hear.

Grab the cyclic stick and collective as your back seater dons his night-vision goggles. Transfer them back when he's ready and snap your gogs onto your helmet. You won't use them until you're done with sensor work.

"Heads down."

Acquire the bike quickly. Watch the man steer the motorbike toward the Marines, stop one hundred yards away, turn around again.

No, no. We can save your daughter. C'mon, man. Trust us.

He repeats the process once more. Unsure.

We're the good guys. We're here to help. We promise.

You wish you could tell the father that Dustoff is almost out of gas. Wish you could encourage him to approach the Marines. Let them save her life. Wish you could say your attack helicopter is for protection. You're on the same side.

The motorbike heads off a third time, disappearing into a sea of square compounds.

"Bingo fuel," the Blackhawk announces.

The JTAC releases him.

Continue in the overhead for another minute, searching for a bike that will never return.

"He's not coming back, is he?" you transmit over your intercom.

Check off station. Turn south toward your next preplanned mission.

It's pointless waiting to save a girl who can't be saved.

Essay Honorable Mention

Michael Farwig

Cursed

I learned how to curse in the Army. How to really curse. How to string together four-letter words and arrange them in combinations that would make infinite paragraphs. How to say only one in a single breath and have it be an entire sentence, a complete conversation.

"Blake's wounded pretty bad."
"Shit..."

Words have power, and curse words maybe the most. You learn this your very first minutes in. A raging Drill Sergeant comes on the bus that brought you from the worn-down Midwestern town you were so desperate to escape to the woods of Missouri, and it seems the only words he says that you can make out are the ones you only whispered back at home out of earshot of parents and teachers. Ones you said on the playground to would-be bullies to try to sound brave, but it always seemed like everyone already knew some truth about you that it would take years for you to find out yourself.

"Lee might not make it."
"God damnit. . ."

You can use the power of the curse word on a ruck march. That's the thing you do the most in the Army. It's not shooting, it's not throwing grenades, it's not hand-to-hand combatives. It's walking. Walking in a line following the person in front of you with a heavy ruck on your back. Walking for miles and miles without ever really knowing where you're going, or where you are at. And it seems like with every march, every year, every deployment, that hike gets a little longer and the pack gets a little heavier. Until it crushes you.

On any given march there comes a point where you think about quitting. The pain in your body seems to be too much. You're tired and thirsty and sore, and you just want to sit down, just for a little bit. But you know that if you sit down that you won't get back up. So, if you curse enough the pain recedes, not all the way, but enough to allow you the space to take another step. Then you say another curse word and take another. A small rhythmic chant ensues, each step forward accompanied by a minor expletive. Until the ruck march ends—or you run out of air.

"MacGowan got shot through the hands."
"Bastards. . ."

My favorite way to curse was in Pashto. That language is made for swearing. Three enthusiastic Afghan police officers sat with me after missions, and we traded words like currency in our own dialects. I remember how their wrinkled faces would crack open into a wide yellow-toothed smile when I'd say how that "fucking" sucked or how the lieutenant was being a "little bitch." I remember thinking how they laughed like little kids with each curse word and how I never knew how old they really were. Their faces were weathered and ancient, but their eyes always gleamed young. There is a chance that these men had learned to curse with a Russian soldier that was my age in the '80s. There is an equal chance that they were younger than me and I was the first to expose them to bilingual vulgarities.

I swapped my English curses for their Pashto ones. They would tell me how it was هنوړغرس or تیاره شوی and I'd quietly say the words myself. They'd laugh at my mispronunciations and try to correct it, and I'd repeat it in the same way that I had before. And they would say whatever curse word they had just learned with harsh and spitting accents. We started our own secret language, intermixing their bad English and my worse Pashto into an incoherent babble of obscenities only we could understand.

"Platoon Sergeant is going to lose his eye."
"Jesus Christ. . ."

I didn't have much time to practice. One month in country and the attack on our base happened. I awoke first to the sound of screams. Then to the sound of bullets and shrapnel ripping through the air, through our tent, through flesh. The power cut out, and the only light was what shone through the holes that were made. I could see through one that it was a clear blue morning.

That first scream was Blake. He was hit in the back and ankle. Before I could even put my body armor on, Lee took shrapnel to the neck. I did a blood sweep on him, checking to see where he was wounded. And even though my hand became warm, wet, and sticky when it touched his throat, I still held it up to one of those light shafts pouring in through the holes in the tent. A dark red blush spread like fire over my palm. I never learned the Pashto word for what I wanted to say then.

Four more members of my squad would be hit in that tent, and I crawled to them and ineffectually bandaged them all and waited for the bullet or shrapnel that would hit me, wondering if I would lie there in that light shaft it made. Who gives first aid to the person who's trained to give first aid?

Minutes, hours, lifetimes passed. And then as quickly as it came it ended. For a brief second, I heard a silence so loud it hurt to listen. It wasn't just the AK 47s and Chinese grenades that halted. The screams of the wounded, the sergeants yelling directions, my own racing heart . . . everything stopped. The rending destruction had created a vacuum, a void so large that not even sound could escape. Then a dying man screamed and it filled that void, filled every empty space and hole, and broke the stillness.

When I walked out of that tent, I looked up and saw that blue sky that had poured through the holes as if it had made them. The sun silhouetted two towering mountains, and the air was clean and dry. Afghanistan always had within it the power to be beautiful if it only tried. Then I looked at the ground and there were five bodies lying in the dirt and dust and rocks. Five lolling heads, ten eyes rolling towards that blue sky in blank vacancy, twenty limbs spread out flailing in all the wrong directions. Five dead, four of them were the enemy. Afghanistan would always be ugly, no matter how hard it tried.

Two of the Afghan police officers that sat and laughed and cursed with me died during that attack. They were not among the bodies around the tents, dying further away. Alone. The other member of the trio died later on in the deployment to an IED blast. I remember telling my interpreter how سرغړونه the whole thing was. He nodded and told me that it was indeed fucked.

"Kolakowski might lose his leg."
"Fuck..."

I didn't curse for a while after I got back from Afghanistan. I didn't really say anything. I'm sure my silence was the most deafening thing to those around me. But what was I supposed to say? Maybe I should've let it go. Maybe I should have screamed and broken something and said how لوړخ the whole thing was, how absolutely سرغړونه I was. Maybe I should've told everyone how angry I was that they died and that it should've been me, and so later on in the deployment, after I got wounded by a roadside bomb, I lay in the hospital bed and wished I was dead. Wished that the explosion, instead of going off just a few feet in front of the vehicle I was driving, had gone off directly under. That instead of merely injuring me, it had done what it was intended to do. That seemed so much better than the alternative that one night, I grabbed my rifle, went into an empty tower on our base, and sat with it under my chin.

The night was cold because I wasn't wearing my body armor, because what would be the point in that. But the barrel was hot against the underside of my jaw, and I could feel the smoothness of the metal there making a deep impression. I sat with it there and stared at the moon, which was crescent and yellow and looked exactly like an Afghan smile I had once known but was already forgetting. I leaned forward and pressed the tip further into me and my fingers found that terrible tension of the trigger. I pressed it down, just to the point where you could feel the pull of it, let my finger sit in that spot between life and death before slowly releasing it. I did this over and over again. I shivered and sat there with the tip pressed against my jaw, until the moon sank and the prayer calls from mosques miles away rang out off the mountains behind me. I eventually put the rifle down, but I came so close so many times.

Because I remembered Lee cursing when that shrapnel hit his neck, and Bowdish cursing when the bullet hit his thigh, and Leininger cursing when he nearly died. I remembered all these curse words being spoken and all the blood soaking into my hands as I tried desperately to bandage wounds that would never heal, and that day I didn't know if I could march on anymore, wasn't sure that cursing would get me to take another step, and it just felt like if I could sit down, even for a little bit, that maybe I could ignore everything that had happened and that it would go away and I could get back to normal, but I couldn't sit down and it didn't go away and nothing was ever going to be fucking normal ever again so you had to just toss the bandages in the trash along with the pain and the hurt and the blood because fuck it right. It doesn't really matter anyways.

> "Tom's dead. So are at least five of the Afghan Police."
>
> ". . ."

And no amount of curse words, no matter how powerful they are, would ever change that.

Tiffany Avery

Fear—Cut Sling Load

When the instructor gave the signal, I stood up and inched backward until I was attached to the Blackhawk only by my grip on the rope and my boots on the metal ledge of the open door.

I moved my gloved hands down the rope, creating more distance between my body and the helicopter. I was grateful I had pulled extra slack into the bird when I boarded.

My body shook as I went further outside the bird. In a moment of courage, I loosened my grip on the ropes. I had so much slack on the line that I fell back a few feet and stopped only when I clamped my hands tightly on the rope.

Startled, I checked my position. I was deeper than ninety degrees, feet now positioned higher than my butt, which was being pushed back and forth in the wind and rotor wash. I swayed towards the front of the helicopter and noticed one of the crew chiefs behind the glass windows. He looked at me through his thick safety goggles and nodded his head. His eyes said, *That's alright, nothing wrong with that.* He gave me a thumbs up.

The hardest part was yet to come—releasing my inner fear and leaping into the wind. This jump would be the culmination of an internal build-up of anxiety, apprehension, and adrenaline, gurgling inside me like a hot gas waiting to explode. The anticipation of the explosion terrified me as much as it thrilled me.

The instructor in the helicopter had warned us not to jump until we had his signal. Failure to obey his commands today meant going home without graduating.

My boots, wet from hours in the dewy grass, slid against the metal ledge. A strong gust of wind came ripping from the tail end of the helicopter. Before I realized what was happening, my feet slipped off the ledge of the bird and I fell outside of the helicopter.

Zero Day—Weeding Out the Weak

The purpose of Zero Day is to weed out the weak. So much of being a Soldier comes down to mental strength and how much heart you have. I might not be the strongest, fastest, or smartest, but I make up for what I lack with heart and my drive to succeed, especially as a female who constantly needs to prove herself in a male-dominated organization.

Zero day began with a two-mile run, after which two Soldiers were already on their way home. Before we caught our breath, the instructors ran the entire formation to the obstacle course. A friend of mine working as support staff took a video of us running, shouting "Air Assault!" every time our left foot hit the ground. He posted the video to Facebook and captioned it "What have YOU done lately?"

Once at the obstacle course, the entire 250-Soldier class looked to the sky and surveyed the swiftly moving clouds, worried about rain. We crowded around the instructors in a gaggle fuck.

"There are three major obstacles, and you cannot fail any of them!" The instructor yelling was short, maybe five foot four, and he was frowning. "There are six minor obstacles, you can fail one! You will have two chances on each obstacle! If you fail a major obstacle on your second try, guess what! You're going to the house! If you fail more than ONE minor obstacle on your second try, you're going to the house!"

Going to the house meant being sent home without graduating.

"When I call your name, go line up at The Tough One!" he shouted. Thunder boomed in the distance. "Abbot, Ackman, Adler, Aikers, Anderson, Archuleta, Atkins, Austin, Avila! Go, hurry up, now! Move!" My heart jumped in my chest.

We sprinted to The Tough One and formed two lines. I was going to be the first female to attempt the obstacle course.

I was scared, I was confident, I feared failure.

Minutes later, it was my turn. "Next two! Let's go!" I turned and faced The Tough One. The first part was the rope climb, notorious for sending people home.

The rope climb was twenty feet. I walked shakily to the rope and took a deep breath. I thought of the rope I had practiced on dozens of times at home. I bought it off Amazon, climbed the huge maple in my yard, and threw it over the highest branch I could reach.

"Ready, go!" shouted the instructor. I grabbed the rope as high up as I could reach and jumped, pulling my knees to my chest. I used my feet to create a lock between my boots and the rope. I straightened my body and reached my hands up again.

In four moves I was up to the wooden beam and propelled myself over. I dropped onto the wooden beams, ran across them, and scaled the triangle ladder. At the top of the ladder I shouted, "Air Assault!" and carefully navigated over another huge wooden beam. My boots reached for the cargo net beneath me and I carefully made my way down.

"Sprint to the next obstacle, go!" yelled an instructor.

I sprinted to The High-Step Over. I placed my hands on the back of my head, interlacing my fingers. I swung my leg and slammed the back of my thigh onto the first wooden beam. I used my grounded foot to boost myself up until I was straddling the beam. I twisted my body, brought my other leg over the beam, and hopped down. One down, at least ten more to go.

Next came The Weaver. The Weaver looks like two huge wooden ladders leaning into each other, meeting at a point fourteen feet in the air. Maintaining three points of contact, the Soldier must move their body over one beam, then under the next, and repeat this pattern up the first ladder and down the second ladder.

When I finished The Weaver the instructor impatiently yelled my name.

I ran over to him and stood at parade rest. "Yes, Air Assault Sergeant!" I knew this instructor because I had deployed with him two years prior.

"Did you yell 'Air Assault' when you reached the top?" he asked. I hesitated. I had *not* shouted at the top of The Weaver. I knew telling the truth would mean repeating The Weaver, which I did not want to do.

"Yes, Air Assault Sergeant!" I yelled. I was now fully committed to this lie. One of the instructors that morning had told us that no matter what, we must fully commit to everything we did at Air Assault school.

"Are you sure? I don't think I heard you." He smirked.

"Yes, I'm sure, Air Assault Sergeant!" I shouted again.

"You're positive?" he asked a third time. Don't falter.

"Yes, I'm positive Air Assault Sergeant!" I tried not to smile, but we both knew I was lying.

"Fine. Get out of my face. Go!" He shook his head and waved me away with his hand.

I failed the next minor obstacle on my first try, The Six Vaults. It was a set of six wooden beams four feet high. Using only your hands, you were supposed to fling your entire body over the beam. After my failed attempt, an instructor yelled at me.

"Have you failed a minor obstacle yet?" he asked.

"No, Air Assault Sergeant!" I looked at the brim of his intimidating black baseball cap.

"Stand off to the side and regroup. When you're ready, go kill it. Now go away!" he said.

A few minutes later I said, "I'm ready, Air Assault Sergeant!"

He walked over to me. "Don't fail."

"Air Assault Sergeant, I will *try* not to fail!"

He cocked his head slightly. I couldn't keep from smiling.

"You're a terrible person, the worst! Absolutely awful!" he grinned.

I finally noticed his sea-blue eyes and well-tanned wrinkles. I wasn't sure what the correct response was, so I said, "Thank you, Air Assault Sergeant!"

He burst out laughing. "That's the attitude, you'll be fine! Now go!"

With that boost of confidence, I promptly failed the obstacle again. It would be the only one I could fail.

The next obstacle was The Incline Wall. It was a flat wooden wall leaning at a forty-five- degree angle, more than six and a half feet high. I jumped and grabbed the top of the wall. I did a straight pull up and swung my leg as hard as I could. My heel caught the wall and I pulled up with my arms. Keeping my head down, I lurched the rest of my body over and down the other side.

Then came the Swing Stop Jump, another major obstacle. I ran at the rope, grabbed it, and swung through the air, aiming my feet at the beam. I barely caught the beam with my heels and nearly fell off. I wobbled until I had my balance, jumped down, and ran to the next obstacle.

The Belly Crawl was the easiest obstacle. Wire was strung sixteen inches off the ground over a sand pit. The goal was to low crawl on your stomach with your head on the ground until you cleared the wire at the other end. When I finished I began to wipe sand out of my eyes.

"Don't wipe that sand off! Wear that sand proudly!" someone yelled.

I immediately dropped my hand from my face and ran to The Confidence Climb, which looks deceivingly easy. The tricky part was the increasing distance between the rungs as you climbed. Once down the other side of the ladder, I ran for the last obstacle.

"That can't be your top speed! My grandma runs faster!"

"Yes, Air Assault Sergeant!" I ran back to The Confidence Climb and sprinted back to the final obstacle.

"Faster!" the instructor screamed. "Go back!"

I raced back. After two more attempts I was winded, panting, and sweating. My heart burned, but I was prepared to run back and forth as many times as he wanted.

"What are you waiting on! Last obstacle! Go!"

I jumped on the low beam of the The Low Belly Over. When I practiced the day before, I couldn't make it over. *Dig deep*, I told myself. I leapt as hard as I could and smashed my stomach onto the beam. I made it over and fell on my butt but immediately jumped up and began running.

An instructor waved me off. "To the barracks!" he said.

I'd made it through the obstacle course.

Back in the barracks I waited. Over the next three hours, females slowly trickled in.

The female lieutenant who had gone to Infantry school and attempted Ranger school, a tall, hard-faced woman with short hair who had dripped confidence, never made it past the rope climb.

The two girls I met when I first arrived at Camp Crowder, the short, muscular CrossFit geeks, didn't make it either.

Another female lieutenant with red hair and anger in her eyes also failed out, calling The High Step-Over bullshit, saying short people had a disadvantage.

Des had made it through, though, the first friend I made here. Small and wiry with short, dark hair, Des was straightforward and funny. We had bonded over our cheap Denny's meal the night before Zero Day.

Henderson, a cadet who didn't look like she had the guts for this, also made it. I admit I had judged her because she came to Air Assault school with only one nametape. When she wore her cover, she had devised a way of wearing her bun high and folding the back of the hat under so no one could tell the nametape was missing. I thought if she hadn't even managed to bring the correct number of nametapes there was no way she was prepared for this course. She proved me wrong.

After Zero Day, we lost over half the females. There were only twelve remaining. I surveyed those left, the girls I would be sleeping, eating, and showering with for the next eleven days.

Those eleven days would be nonstop training, including a six-mile ruck march, a four-mile run, three major written tests, four strenuous hands-on tests, five rappels including a ninety-foot rappel out of a UH-60 Blackhawk, and the final twelve-mile ruck march.

As we got ready for bed that night, Des smiled at me and put her fist up. I bumped her fist, returning the smile. "Let's do this," she said.

Phase One—Aircraft Orientation (The Easy Phase)

Phase One was Aircraft Orientation. One of the many requirements for this phase was learning all the specifications for nine different military aircraft. We also learned Hand and Arm Signals, which was my favorite part. We learned how to direct an aircraft to take off, to back up, or come forward, and how to tell the pilot to set down their sling load. This signal is called "Cut Sling Load."

Phase One ended with a written test and a Hand and Arm Signals test. No one was expected to fail. However, roster number 108, who stood to my left in formation, managed to fail.

Roster number 108 was Austin. He was a 22-year-old Specialist who had been in the National Guard for three years. He was our comic relief.

One day on a quick break someone was taking too long in a porta potty. Our break was short and the lines were long, so someone yelled for the person inside to hurry up.

"I'm taking a crap!" came the reply.

Austin, quick of wit only when it didn't matter, replied from another line, "Cut sling load, cut sling load, cut sling load!" Everyone laughed, and from then on "Cut sling load" unofficially meant "pinch it off and get out!"

Another morning Austin unabashedly described making eye contact with another male while using the toilet. All the males with last names A through D were assigned to the WWII-era barracks. The most charming feature of these barracks were the toilets that faced each other. There were no privacy dividers. Austin liked to joke about his proximity to other males while doing his business, pants around his ankles.

Austin made jokes while we were getting smoked, we had to prod him to stay awake in class, and though he had a decent intelligence level, he got easily flustered and made dumb mistakes. Somehow, Austin even managed to mess up our morning count.

The first thing we did each morning when the instructors came out was count off. The numbering of the ranks began with 100. This meant when the count off came to me, even though I was roster number 107, I was the eighth soldier to count off.

The morning of the first test began with a count off as usual. After roster 106, who stood to my right, sounded off with "Zero-Seven!" I yelled "Zero-Eight!" Austin, roster number 108 to my left, opened his mouth and yelled "Zero Ni—wait, something's not right!" He looked at me.

I looked at his confused face, his forehead scrunched. It had taken four mornings of counting off before he realized his count-off number didn't match his roster number. Roster number 109, a first lieutenant, quickly straightened him out.

"You're the ninth Soldier, you idiot," he told him; "it's not your roster number." I couldn't help myself; I burst out laughing. All the Soldiers close enough to hear the exchange laughed too.

I heard later that Austin failed the test on purpose; he didn't have the heart for Air Assault school. Probably the best thing—if you didn't really want the wings, you didn't deserve them.

When roster number 108 failed the Phase One test a second time, he was gone before we could say goodbye.

Phase Two—Sling Load Operations (The Difficult Phase)

Phase Two covered Sling Load Operations. Sling loads are anything an aircraft can lift with cargo hooks and fly to another place. A huge variety of stuff can be airlifted, from fuel drums to multiple Humvees at once. Being able to carefully inspect these sling loads was crucial.

For the hands-on test, we would inspect four sling loads. Each inspection had a two-minute time limit, and we had to find at least three out of the four deficiencies in each load. For the A22 Cargo Bag alone, there were no fewer than sixty possible places for a deficiency.

The morning of the Sling Load test I was the most nervous I had ever been about any test. Des had spent extra time with me the night before practicing our inspection sequences. I admired and envied her easy confidence.

"Dude you got this, quit worrying so much." She smiled. I wondered how she seemed so relaxed. "And if you don't, you'll know you did everything you could."

I took that to heart. If I failed, it would not be for lack of trying. I decided I could not let my fear keep me from my success.

I was in the last group of people to do the Sling Load test. The instructor grading me was the Air Assault Sergeant I had deployed with.

"When you're ready," he said.

I breathed deeply, flattening down my erratically jumping nerves with each exhalation.

"You know, you're the last person I'll ever grade. This is my last cycle as an instructor."

"Really?" I asked.

"Yep. Make it good. Are you ready yet?" he asked.

"Ready," I said.

"Then go," he said.

I turned around. "Deficiency!" I yelled immediately. The aircraft listed could not carry the sling load in front of me. This gave me confidence and I moved on quickly. It took me less than a minute to find all four deficiencies.

"Next!" yelled the Air Assault Sergeant.

I ran to the next load, the A22 Cargo Bag. I followed the steps carefully. In ninety seconds I found all four deficiencies.

"You're on a roll," the instructor said.

I smiled and ran to the next sling load, the Humvee. I must have let his compliment go to my head. I only found three deficiencies before I ran out of time.

The last sling load was the water trailer. The easiest load of them all; you could see almost everything when standing at the hitch. By the time I had worked my way around the trailer, I had only found two deficiencies. I started over, beginning to panic. I took a step back. *They don't exist*, my brain said.

"Twenty seconds," the instructor warned.

I looked at the trailer, searching frantically.

The instructor began to count down. "Ten, nine, eight, seven—"

"Deficiency!" I yelled. I threw a knife hand at one of the tie-downs. It was lacing cord, and it was supposed to be Type III Nylon. Such a simple error.

I exhaled and smiled. The Sling Load test was over.

Since I was in the last group to test, by the time I returned to our barracks I was hoping to see all the other females in the bay. As I walked around, I realized five females were still missing. I was glad to see Des, my bunk neighbor, eating a chili mac MRE on her rack.

Those of us who had made it through waited anxiously. The retest was that same day. We hadn't lost any other females since Zero Day. Together we had endured icy cold showers, bruises that looked like we'd been kicked by horses, two tornado drills, and moving barracks three times. We were bonded and all rooting for each other.

When the remaining five girls trickled in, our excitement filled the bay. All we had left was rappelling and the twelve-mile ruck.

Phase Three—Rappelling (The "Fun" Phase)

Rappelling was supposed to be the "fun" phase. We learned to tie hip rappel seats, completed five rappels at increasing heights, and finally, arrived at helicopter rappel day. Likely the only time any of us would ever have the chance to rappel out of a Blackhawk. For some, it was their entire motivation behind attending Air Assault school. I was terrified.

Part of overcoming fear is pushing aside those who don't believe in you and using their negativity as the accelerant to light your fire. I had doubters back at my unit, including a surly readiness sergeant and a second lieutenant who had scoffed to me, "They're sending a *cook* to Air Assault?" I wanted wings to show myself I could do it, but I also wanted wings to show the doubters they couldn't impose limitations on me. When I jumped out of the Blackhawk, I would think of the doubters.

On Blackhawk rappel day, the pilots said it was too cloudy to fly. We were already at the landing zone, so had no choice but to wait in the cold for the cloud cover to clear.

Two hours later, the instructors received word the birds were on their way. We tied our hip rappel seats, yanking the loose ends of the rope tighter, squatting deeply, and forcing the rope to dig into our skin. Then the instructors gave the signal to line up for inspection. The group ran like hell to the instructors. First to pass inspection would be first in line to rappel.

I ended up in the back of the pack. The worst part would be the anticipation. I wanted to jump, desperately; also, I didn't want to jump, not even a little. It wasn't the height I was afraid of—I was afraid of failure.

We had already been outside for hours. I could feel my body temperature dropping. The temperature remained in the fifties, it was extremely windy, and raindrops flew around, assaulting our bare cheeks and noses. I tried to figure out how long I would be waiting for my turn.

The two Blackhawks each take four rappelers at a time. There are 215 Soldiers waiting. When the first group of Soldiers ran to the bird, I began timing.

I watched as the first group of Soldiers scrambled into the bird and got on their knees. After safety checks, the birds lifted off the ground.

That part took two minutes. The Blackhawks climbed into the sky until they reached their hovering distance of ninety feet. I watched as the Soldiers inside jumped out and rappelled to the ground.

Once the Blackhawks touched back down, I stopped timing. The whole process took less than six minutes. I did a rough mental calculation. It would be almost three hours until my turn. Uncertainty is hard to deal with; certainty, even when it sucks, is a little bit easier to handle.

Then I saw one of the females run to a Blackhawk. I could tell it was Henderson because of the tufts of unruly blond bun sticking out from under her helmet. She told me she was terrified of heights. She didn't want this jump to happen at all. I watched as she went up in the Blackhawk, praying she maintained enough confidence to make it through.

I saw the instructor in the bird give Henderson the signal to get ready. I watched her trying to work herself down the rappel line to get her body in position. The two rappelers on the other side jumped. Henderson didn't move. The person next to her jumped too. Then the instructor inside the bird grabbed Henderson and yanked her back inside the Blackhawk. *Shit*, I thought. *She must have lost nerve.* Refusing to jump means not graduating. I didn't want that for her.

I continued waiting, slowly inching up in line. I shook with cold, unable to feel my fingers. I wondered if that would affect my ability to hold the rappel line.

More than two hours later, I was next in line. My heart rate quickened and I felt my pulse thrum in my ears. I watched anxiously as the helicopter touched the ground. I hunched over, keeping my profile low, ready to run to the bird.

One of the Soldiers working as safety ran over to me and the other Soldier waiting. "They have to refuel," one of them yelled. "They said to go to your stuff!"

Fuck! I thought. I watched both Blackhawks fly away.

I walked away. No one knew what was going on. I found Henderson and asked her what happened. I liked Henderson because she has the same kind of determination I do; the kind that gets boosted by doubters and the need to prove herself as a highly capable woman in the Army.

"You don't even realize how hard it is to pull the rope in when you're up there. There's like no slack. I don't know if the safeties were holding it too tight, or I wasn't strong enough, but I couldn't get any slack," she said. She was talking fast, hyped up. "I couldn't lean out because I couldn't get any rope. The instructor tried too but he couldn't get any," she told me. "At first he thought I didn't want to jump. But when he couldn't get any slack, he just like, threw me on the floor of the Blackhawk."

"Shit," I said. "Are they going to let you go again?"

"Yeah. He knows it wasn't my fault. The other instructors on the ground called me a pussy and weak, and said I'm going home." She took a long drink from her canteen. "They said I can go again, but I have to get in line at the end."

"Ugh, I'm at the end too. Are we going to get to go?"

"I dunno," she said, shrugging her shoulders. "I hope so. Now that I've been up once, I'm ready!"

Next I found Des. She had already jumped. "How was it?" I asked.

"Fucking amazing dude!" She had a big smile on her face and her eyes were wild. She was one who had been looking forward to today. Her excitement made me feel a bit better.

Forty-five minutes later, we heard the birds in the distance. We turned our eyes to the sky. The Blackhawks were back.

The 12-Mile Ruck—The Final Test of Endurance (and Heart)

The twelve-mile ruck was the least of my difficulties. For me, a distance like this was an individual event. I wasn't spurred by the words of encouragement from the people around me. What got me going was my own determination to always improve. I relied on my mantras when it got hard; *dig deep* and *I want this*. Those five little words kept my fire lit.

In the darkness, once I escaped the packs of Soldiers, I almost felt alone. I had planned to stop jogging when I got tired, or as soon as I was decently ahead of the time hack, but I never felt like stopping.

I crossed the finish line in two hours and thirty-five minutes. I lined up next to the other finishers already emptying their rucksacks for inspection.

"I was hoping to make top ten!" said the male who came in after me.

"Oh really? What are you?" I asked.

"I'm twelfth, which means you're eleventh," he smiled at me. His head was soaked with sweat, making his silver hairs glisten. He had a stocky build and a slightly weathered face and his eyes were a light, bright blue. "But I'm old, too."

"Oh yeah, how old are you?" I asked.

"Thirty-six," he said. "I'm an old guy."

I smiled, a little too pleased. "I'm thirty-seven."

"No way!" he laughed. "Really?"

"Yes," I laughed too.

"Now I feel worse!" he said. He smiled and shook his head. I could tell he was thinking, *Beat by a female, and an older female!*

After the inspection we went back to the barracks to get cleaned up for graduation. I hobbled to my bunk and saw Des sitting on hers.

"I didn't make it, dude," she said. Her eyes were tired and dark.

"What! What happened?" I asked. My heart tightened.

"I don't know. I was just starting the last mile, staying ahead of the trail vehicle, and they said I missed the time hack and there was no way I could make it." I felt her disappointment. As happy as I had been, Des not making it diminished my joy.

I felt guilty for not staying with her during the ruck.

She shook her head as if she'd read my mind. "It's okay, man. My fault."

I didn't know what to say. After everything we'd been through, she was out at the 11th mile.

As I showered, I experienced mixed emotions. I was immensely proud of myself and all the other females who made it. But I was devastated for Des and the two other females who fell out of the ruck and wouldn't graduate. I felt guilty, I felt relieved, I felt happy, I felt pain, physically and mentally.

Blackhawk Day

I dangled outside the helicopter, my left arm throbbing from smashing into the side of the bird when I fell. My arm was now trapped tightly between my body and the metal ledge.

"I slipped!" I screamed, "I slipped!" I was afraid the instructor would think I had jumped without his signal. "I slipped!"

The instructor turned around and looked at me, legs dangling in the air, arm smushed tightly against the bird. He crawled over to me on his knees and leaned out the side of the helicopter.

"It's okay," he yelled, getting as close to my face as he could. "Move your arm down and go!" He patted my right arm, still extended and gripping the rope above my head.

I nodded my head. It was too late to do a four-point body check, but I turned my head and looked to the ground. I released my grip.

As I rappelled down the rope, I smiled. I didn't think about my fears or the doubters even once. They had slipped right out of my mind.

Billy Jenkins

The Night Before

I sit on a sand-dusted cot, cleaning my rifle in a crowded tent, surrounded by 90 or so other Marines in the Kuwaiti desert. It's just before evening chow, rumor has it we push north tomorrow, invade the country we've come here to wage war against. I am scared. Period point blank. I feel as though, with the rush of excitement, I've been caught up in the wave of patriotism that's swept across the purple mountains majesty. The amount of time between me holding a remote control in the armchair of my living room while watching CNN to sitting on this cot in this tent feels like no time at all.

As far back as I could remember, I wanted to be a Marine. My father was a Marine and he served back in the 50s. When I was a kid, he'd wake me up and carry me downstairs so we could watch old war movies together. We'd make machine guns out of our fingers, shoot each other from behind couches, and throw balled-up paper grenades at my mom and sisters. He sang marching cadences in the car and told story after story about his time in the Corps. I grew up whistling the theme from The Bridge On The River Kwai, *and watching Jim Brown make his last minute, always-fateful, run to freedom in* The Dirty Dozen. *Eating cold beans off a metal plate looked cool because The Duke did it; sleeping in a foxhole under the stars would be tolerable because Charles Bronson never complained when he did it on TV. I was a kid then, I'm not a kid anymore.*

While I sit on this cot, while the Marines in this tent busy themselves, I think about how I got here.

"Uh . . . roger that," I said into the receiver as I hung up the phone. I sat on the edge of the bathtub. I placed the phone on the sink and leaned back against the cold white tile. A bright fluorescent light hummed on the wall in front of me. I stood, taking stock of myself in the mirror. I put the phone in my pocket and left the bathroom; I paused at the top of the darkened staircase. My daughter Emily and niece Marissa, sat in the living room below; I heard cartoon laughter coming from the TV. It was a Saturday evening, the week before Christmas, and the house was bathed in the glow of slow-flashing lights. Stockings lined the bannister, decorations covered every piece of furniture. I listened to Emily and Marissa, and beyond them the sound of my mother and sisters talking in the kitchen. Quietly I moved downstairs.

"Hi, Uncle B."

"Hi, Marissa."

"Hi, Daddy."

"Turtle," I said, as I leaned in to kiss her cheek. "I love you very much."

"I love you, too, Daddy."

I tickled Marissa as I passed her and made my way through the living room. A thick, heavy snow fell outside the picture window in the dining room. Cars, already half-covered, looked like sleeping babies under their fresh white blankets. I exhaled and stepped into the kitchen. My mother held court at the far end of the table, while my older sister, Barbie, and my twin sister, Theresa, flanked her on either side of the table.

I paused for a minute and smiled. I took a seat at the end of the table and looked at each of them; I cleared my throat.

"What did you just say?" my mother asked.

Before I could answer, the kids, laughing, skipped into the kitchen. Marissa ran to her mother while Emily hugged me and climbed into my lap. Both started asking for more cookies.

"I said, I have to go to Iraq."

There would be no more waiting to see what our president, our military, or the country was going to do post September 11, 2001. The time was upon us to invade Iraq. It had been eleven long years since I went to boot camp, since I'd graduated from Parris Island, and now I would be put to the test in the crucible of combat. That's a long time to wait, a long time to prepare. To wonder and not know. People used to ask me if I really wanted to go to war, if I really wanted to fight. I used to liken it to someone who studies for years to be a doctor, or a lawyer; a mechanic or an athlete who practices every day for a game. Of course, I wanted to go, or rather, of course I thought I did. I had the luxury of not being this close before.

When we shipped out, I was scared I would fail as a leader of Marines. I had been a platoon sergeant at my reserve unit, which is to say, I oversaw thirty men, professionally and personally. Teaching them what war might be like. Everything from firing a rifle to packing the right gear. It all seemed so foreign now, so distant. We were thirty miles from Iraq, and all those years of training seemed wasted. Had I taught them how to survive? Had I given them the knowledge they would need to make it home? Questions. I was sick with questions about my abilities. Maybe all soldiers headed to war are.

Back in Pittsburgh, it was the faces of the parents looking down at me with quiet panic in their eyes at the thought of never seeing their sons and daughters again that I think of now. Their looks of half-reassurance at my qualifications as their child's boss as we boarded our buses is a vice grip of pressure I haven't stopped feeling in my gut. When we left home on that chilly January night, I pretended it was no big deal to my family. I made home videos with them. I brushed my daughter Emily's hair. My mother and sisters cried. I tried to maintain a cool exterior. An exterior that I must rebuild here every day, a façade of calmness, while war waits for us across the border. Waiting, like a summer thunderstorm on some distant horizon.

While I wipe and re-wipe the same spot on my rifle, I sit here and keep replaying the last few weeks in my mind. I try to organize everything we have done so far like I'm packing my gear. There is a reassurance in the packing and repacking of gear. That sense of completion that checking things off a list brings. I repack my thoughts carefully.

We were sent to Camp Lejeune N.C. for about a week before we left the states. We were issued more gear than one person could possibly carry, and given enough shots in the arm to qualify us as bona-fide science experiments. The anthrax pustule that has lingered unpromisingly on my shoulder, which was meant to counteract any biological weapons Ol' Saddam had in store for us, just seemed to sit there bubbly and sensitive and ominous.

"What exactly did they pump into our arms?" was the question I heard most around the barracks in North Carolina.

Our time there was filled with busy work, dental appointments, and medical screenings; I suppose the military felt it was important not to send someone into combat with a cavity or with athlete's foot. We tried not to think about what might happen to us once we got in country, once the shooting started. In the military, as in life, or at least on all those old black and white war movies, it was always the other guy who got "it," always the other guy who didn't make it home. I continuously looked around my unit hoping to spot those less prepared than me, hoping to find the Well-At-Least-I'm-Better-Than-He-Was kind of Marine.

When our time in Carolina was over, we boarded our jet and headed to Kuwait. I mostly thought about my list of lasts; my last drive around Pittsburgh, my last look around my childhood bedroom, my last Christmas cookie, my last Will and Testament, my last everything with my daughter.

When we stepped off the last plane in Kuwait, it was like getting smacked in the mouth with an open preheating oven. My God. The heat of Kuwait was beyond intense. My company, Military Police Company Bravo, was then bussed to a newly built tent city called Camp Bougainville, which roughly 1,000 Jarheads would call home before the invasion.

Our tent, built for fifty Marines, houses ninety, and it reeks of sweat and tobacco and gunpowder. There are cots against each wall and a row down the middle. Sea-bags and flak-jackets lay all over. We are crammed inside this tent like sardines, like inmates in an overcrowded prison. Marines tied rope across the tent and hung their trousers and t-shirts and underwear on it. Sweat socks full of water bottles pepper the tent wall. At times, it looks more like a refugee camp than military housing.

I've learned that the desert is big, and yet I can't seem to take a step without tripping over another Marine from Bravo Company. My cot rests between the tent wall and another Marine's bunk; I claimed this parcel of land because it seemed better than being sandwiched between two grown men for God knows how long. My cot is a green canvas and aluminum framed respite from the long days on my feet. I think it was built by Oompa-Loompas because it is neither long enough to lie on without my feet dangling over, nor comfortable enough to get any real REM sleep. It is, however the perfect height and length for slamming my shins into daily. Lying beside me, mere inches from my face, is Eddie Paluch, an outspoken kid from Buffalo. At night, we can feel each other breathe.

I sleep with my back to him, which has given the tent wall the opportunity to pound me about the face and neck when the evening winds kick up. By the third day, I became a stomach sleeper.

The camp itself houses the collective weapons, vehicles, and Marines of the ever-growing fighting force. Generators hum obnoxiously, while the ever-present aroma of diesel fuel permeates my nostrils.

During this time at Camp Bougainville, my unit and every other unit on base is trying to cram each minute of daylight with classes chock-filled with information we're supposed to never forget. Classes on proper desert hygiene and water consumption, periods of instruction on field-expedient trauma dressings, and how to call in a medevac chopper. The frayed green notebook I carry with me always looks like a generations-old family cookbook, with rabbit-eared pages and circled entries. My indecipherable chicken-scratch will be the death of me if I ever actually need to use the damn thing. Each night I return to the page entitled "Key Arabic Phrases!!", which is pretty much useless because I can barely make out the haphazard way I scribbled down "Stop!" and "Put down that rocket launcher!" underneath various pictures of the Arabic alphabet.

Without a doubt, the class on biological weapons insignia keeps me up at night the most. The all-too-real fear of coming face-to-face with one of those signs in my immediate future is a thought that makes me shudder. When my junior Marines ask if I think we will see any biological weapons, beating around the bush like they do, kicking rocks and pretending to be aloof, I lie and say no. I tell them to go repack their gear for the umpteenth time and clean their weapons. The fact is, I hope we don't.

The worst part of the day by far is noon, when the sun positions itself directly overhead and seems to reach down and grab me by the shoulders. At noon, being inside this tent is as close to unbearable as one might imagine, the relentless heat forces us to eat our prepackaged meals on a hillside that runs the length of the camp. We convene behind our tent regularly for this midday chow. Our second day here, Eddie and I perched ourselves on the hillside and started in on lunch.

"What did you get?"

"The five-fingers of death, you?"

"Fucking beef stew again. What the fuck!"

As we began, we felt the whisper of moisture being carried on the slightest of breezes. Chow seemed tolerable, until I looked up and saw the shit-removal truck hosing down a pair of Porta-Johns upwind from us. The truck became to be known as *Pooh-bear.*

Pooh-bear and the Arabian Christopher Robin who manned the behemoth truck began to represent the dichotomy of our situation. The ever-shifting balance I face between "Keep going" and "I fucking quit." This Pooh-bear driver, clad in black, knee-high, rubber work boots, presumably swimming in feet sweat, a raggedy torn kerchief wrapped around his nose and mouth, barely shielding his senses from the wretched and foul aroma wafting up from the fecal material of a thousand scared Marines, his "fuck-it I quit meter" pegging out in the red. Enter his relief, the knowledge that he will stay here, stay safe once the war begins, while every other swinging dick in this camp will be heading north to futures unknown. Suddenly the Pooh-bear driver's job seems kind of enviable. The idea that things could always be worse begins to take root in my brain-housing-group. I remind myself to be thankful for what I have here, now, because noon-chow-Pooh-bear-mist in Kuwait is still in Kuwait.

My days up till now have been like this. I wake up, get classes on various ways to kill people or avoid being killed by people, train under the unrelenting sun, eat some Middle-Eastern version of American fare in a crowded chow hall, then try to snatch a few hours of sleep between the smells of an overcrowded hooch and the snores of Eddie only inches from my face.

Nights here have been better than the days, but not by much. The drop in temperature raises spirits enough to make a difference. Night isn't the same as sleep, and sleep is a luxury I have learned to forget about. At night, my thoughts drift to our neighbors to the north. The Republican Guard. Saddam. What combat will be like. I lie awake in my cot most nights exhausted by the day's routine, the day's monotony, drifting down the river of self-doubt, my raft overflowing with second guesses and questions from my junior Marines, questions I can't answer. Our already jam-packed tent of men is made more crowded by our unspoken fears, which hang in the air like a heavy fog.

Some nights my platoon draws guard duty, and we patrol the camp in a Humvee. I like this duty, there's a solitude to it, a time to data dump the day's events. Eddie and I usually work it together, and we take our vehicle up on a tiny hill behind the base and survey the sleeping masses below. You can see Iraq in the distance, oil wells that burned in the night. The smoke creates a haze that will trap the rising sun's brilliant palette of colors; their momentary magnificence rivals any sunrise on planet earth.

Somewhere between night and day, between navigating Pooh-bear visits, between classes on nerve agents and medivac procedures, somewhere before dawn, before the sounds of angry leaders yelling at fresh-faced troops replaces the stillness, this place is calm.

No matter the time of day, the fear remains. The worry I carry for my Marines continues to gnaw at me. I need to let them learn these classes on their own, learn to take care of themselves. I can't micromanage them. This of course is one of the hardest things for me to do. The time it takes showing the juniors how to do a thing, load a machine gun, dress a pretend wound, fix a broken radio, then letting them fail at it, so they can remember on their own is time I don't have. The time-to-combat-clock is ticking. I want free-thinking problem-solvers, not robots, but the time it takes to mold a free-thinking problem-solver is time I don't have. Trying to replace the nervousness they feel with confidence in tasks completed is a tall order. It knots me up inside and swirls around my stomach. My only defense against their fear is the poker face I have come to master over the years. Reflecting a look somewhere between stoicism and rage, it scares the living shit out of my junior Marines. Whenever their questions start in, I must mask my complete and utter ignorance about our fates with my always-ready poker face. It only takes one Marine asking a relatively benign question for the rest to start in with an avalanche of doubt. Invariably this will occur when they have me cornered in line outside chow hall, or in the showers, or at night in the tent when they know I'm just pretending to be asleep.

"So, when are we leaving, Sergeant?"

"What do you think it will be like, Sergeant?"

"Do you think there are really a million Republican Guard up there, Sergeant J?"

"How long do you think we will be here?"

"Sergeant J, will I ever have to call in air support?"

"You're going to be with us the whole time, right?"

Enter my poker face. I'll slowly light a cigarette and tell them to calm down, then just redirect them to some other task. It is impossible for me to let them know I have no idea, that I am just as scared as they are, that once we cross the border they all, we all, may die. I need them to feel my confidence, soak it in, and take comfort in it. I just hope I can avoid any mirrors because I'm not sure I would believe my poker face right now.

These thoughts swirl in my head while I sit on this cot and clean my rifle. Eddie walks into the tent and quickly makes his way over to me. He paces around the crowded space and stares at me pensively.

"Did you know they're serving pizza and soda tonight?"

"No, I didn't," I say, "but that can only mean one thing."

"Yeah, I know," he mumbles.

"Well, let's get some fucking pizza, I guess." And with that we walk in silence to the chow hall for our last meal in Kuwait.

I return from evening chow. The tent hums with restless Marines. I lie on my cot trying not to think about what is about to happen, trying to digest the cheese and pepperoni that sits in my gut like mortar-less bricks. The light from a dozen red-lens flashlights crisscross the hooch. The entire camp is in blackout, probably the entire country, and my tent like so many others is crowded around a tiny radio with the promise of a presidential address. I lie in my cot because I don't need to hear what I know is coming at first light, the invasion.

Before dawn my platoon loads up our vehicles, and we drive in silence the thirty or so miles it is to the border. Hundreds and hundreds of vehicles as far as the eye can see, sit idling, the smell of diesel fuel permeates everything. We stage our vehicles and I walk the line; I nod to my Marines as I pass their vehicles. Most sleep, some watch and listen to the rat-a-tat-tat of machine-gun fire in the distance. Jets and helicopters scream overhead; they drop their payloads on a little piece of land just across the border called Safwan Hill. This border town, only a few kilometers from us, is an Iraqi military observation post. We bomb it further back to the Stone Age. We drop so much firepower on the hill that the haze of smoke leaves only the slightest of outlines. It looks like the sun fell from the sky and landed on this hill. The thunderous report of American bombs and missiles seems endless.

I continue to walk and find Fitz, a young Marine from the slopes of Pittsburgh's Southside. His thick Yinzer accent calls out to me.

"Sergeant J, are you fucking seeing this? The Air Force is blowing the shit out of that place."

"Yeah, they are," is all I can muster.

Fitz somehow managed to commandeer a satellite phone and pulls it from his pocket. He asks if I want to make a call home.

"Jesus, Fitz, how the fuck did you manage that?"

"You know me . . . anyway everyone else is asleep so . . . you want it?"

"Fuck yes!" I snatch the phone and dial furiously. After an endless number of clicks and echoes, I reach my mom. Fitz and I trade off making calls; I talk to my sisters and my daughter. Their voices are warm and familiar. We finish calling as many people as we can and sit atop his Humvee. We stare off into the night and listen to the explosions. Fitz and I can't seem to wrap our minds around talking to our families while standing at the border of a country we are about to invade. We decide we will keep this moment to ourselves.

I return to my vehicle and wait. Hours go by, then our radio clicks a barely readable message about moving out. I take my orders and follow the vehicle ahead of me. We pass through a giant barbed-wire fence and into enemy territory. We drive at a snail's pace; we maneuver around burned-out Iraqi cars and blown-out enemy tanks. The smoke is thick and hovers above the countryside. My vehicle crew is silent. We stare blankly at the carnage we pass. Bodies line the streets as we roll across the border and drive deeper into this invaded country. Arms and legs lay about. Pockets of fire consume half standing structures. In the distance machine-gun fire slashes through the night. In the distance women and children cry.

I drive the Humvee, and simultaneously I watch myself drive the Humvee. My mind floats above me, trying to reconcile what my eyes take in. I am suddenly the Pooh-bear driver; I feel lucky, that I am sitting in this vehicle alive and not one of the dead we pass. I already miss Pooh-bear mist.

We roll through another town. The infantry engages the enemy up ahead of us. Rockets streak across the horizon, flares illuminate the early morning sky, a mosque blares the morning call to prayers.

We have crossed the border and are still alive. I flex the muscles in my feet, and in my ankles, I flex the muscles of my thighs and of my calves, I wiggle my toes. I do this for fear that it will be my last chance to do so. Along with the gear I am issued, I carry with me a measureless weight of expectation and fear. Burdens I cannot shake. I carry a fear of landmines and sniper fire, the fear of being blown up, taken prisoner, of being shot. I carry the fear that one of these things will happen to one of my Marines. Each fear cycles through my mind repeatedly, on an endless loop. I drive on, while the sun rises on my first day in country.

Troy Allan

An Ordinary Wednesday

God speaks to each of us as he makes us,
Then walks with us silently out of the night. . .
Let everything happen to you: beauty and terror.
Just keep going. No feeling is final.
—Rainer Maria Rilke, *Rilke's Book of Hours: Love Poems to God*

A white-haired woman was sitting by herself in the middle of the coffee shop. Her hands were trembling as she pressed her iPhone to place a call. Her eyes were tight with worry, and her teeth nibbled on her lower lip. She had just sat down with a cup of hot tea and a chocolate muffin when four armed policemen ran by the front window. They were forcing ammunition into their assault rifles. They had donned their flack-jackets and helmets and were focusing their attention on the bank across the street, taking up positions behind parked cars and screaming at people to get off the streets.

In 2018, over three hundred mass shootings were reported in the United States. Three weeks ago, the Tree of Life Synagogue in Pittsburg was a target. That tragedy took the lives of eleven people in the midst of worshiping God. The Jewish organizations said the violence at the Tree of Life Synagogue underscores and highlights the dangers of unchecked hatred that runs amuck in our society, and we will continue to see these kinds of violent acts. I am not a stranger to the fear and worry that violence produces, having served two tours of duty to the Middle East and Afghanistan.

I remember one particular day at Forward Operating Base (FOB) Salerno, "Rocket City," on the southeast border of Afghanistan and Pakistan near the city of Khost. The base had recently been the target of a tremendous attack. Insurgents drove a truck packed with 1,500 pounds of explosives into the perimeter gate; a suicide bomber then detonated the truck bomb, causing substantial damage to the base's buildings and security structure. After the explosion, armed insurgents entered the compound with assault rifles, grenades, and suicide vests. The fighters looked drunk as they staggered through the smoking gate and crater that the massive explosion had made. The vigilant fighters took heavy losses to the security forces within the compound. Several insurgents exploded their vests and grenades, leaving body parts and pieces scattered on the ground. As if in some demented rainstorm of blood and flesh, the base was in complete disarray. Imagine cleaning up after that. After the attack, I was immediately sent to the base to provide care for the wounded. I was assigned as a Combat Stress Control Chaplain, and my team and I were experts in care. We were on the ground within twenty-four hours.

The base, the soldiers, and the security forces were a mess. The scale of the attack had shaken the confidence of those that worked at FOB Salerno, although the base had not lost anyone to the attack. Soldiers were worried, scared, emotionally and spiritually injured. My chaplain friend told me that all he can see in his mind is the fingers he picked off the ground, placing them in a trash bag as if it was the normal thing to do, as if he were picking up trash off the freeway on a weekend service project. "My God," he said. "What a nightmare."

Imagine, if you will, sitting in the makeshift dining facility (DFAC), listening to the stories of the attack, the burnt buildings smelling of death, when suddenly another large explosion takes place just outside your location, and the alert sounds. The entire room shook, and then, as if in slow motion, everyone hid under tables or ran outside to the bunkers. It was a living nightmare. My assistant took me by the shirt outside to safety. I remember looking back at the DFAC, empty tables with hot food still on trays lying on the tables. The rocket had just missed us, and I spent the next two hours inside a cement bunker with fifty other people. Not a word was spoken. I missed home, and I missed my wife and my child. I thought about them, sleeping in their beds, warm and safe. Like a wild dog, the reality of war and death was licking me in the face. As a chaplain, I have seen the darkest of humanity in action. The woman trembling here in the coffee shop with the police aiming their weapons across the street reminds me of those I have held, picked up off the floor, and glued back together.

When the speaker turned up on her phone, I could hear the ringing of the other line. The woman waits. She shifts in her chair. With one eye on the street and one eye on the woman, I wonder who will pick up the line. I wonder about the conversation. I am sitting at a table next to the window. The woman and I have a front-row seat to the drama that is unfolding. As the police continue to move people off the street, an older man pushes through the front door and comes towards me. He is dressed in a grungy grey sweater. His pants are tan, with large dark stains. His shoes are worn. His hair is grey. He is only partially shaven and exudes a pungent odor. He seems like someone that lives on the streets; he sits at my table and glances out the window. It seems he was drawn to me, and I assume it was my seemingly calm presence that brought him near. He looks at me with his worn, pale blue eyes and speaks.

"Do you think . . . we will die? The police said an active shooter is in the building across the street."

I wait.

"Do you think the glass is bulletproof?" He taps on the window. "Should we move away from the glass? I don't know what to do. I am scared."

It was as if I could touch the man's anxiety. It shot through him like electricity. Before I could say a word, he stood up and ran back into the street, screaming:

"Put your Goddamn guns down; I don't want to die!"

The tension on the street and in the coffee shop jumped to a new level. It felt like I was being choked, but I tried to remain calm.

It was then, through the yelling of the people outside, that I heard the voice on the other end of the line, "Mom . . . is everything okay?"

There was a pause.

"I wanted to call you," the old woman explained, "because . . . I want you to know how much I love you, and if I don't make it out of here alive, please, please don't ever forget me."

From that moment, the old woman and I became partners on this journey. She was no longer the old white-haired woman in the coffee shop. She, perhaps like me, began to question what it is like to be human. No doubt, there are a million different ways to address the existential questions played out. And, depending on our worldviews and the religious container that carries these thoughts, there can be just as many answers as questions. In other words, with all of these thoughts spinning in my mind, catching every possible wave, it becomes exhausting to imagine my end, the end of my family and friends, and the end of strangers. It makes me question where God is in all of this. Just the other night, I found myself sitting in the darkness, with my naked legs dangling off the bed, and thought, "I can't take another step towards God. I don't want to think about God. I don't want to talk about God. I don't care where I go when I die. I just want to be happy in this life and stop worrying about all of this." I realize the journey towards spiritual maturity, as David Elkins describes it, is not simply an "intellectual journey" but one that "involves the heart." And, the work of the heart is often terribly painful.

Like motion sickness, the sense of sorrow is not always present. There are days when I am incredibly connected to God and feel rooted in his arms, but it is never consistent. Some would contend this is the darkness that enters the soul on a journey to find God—the journey to the center of the soul, where like a glimmering diamond, we find God. It is, or might be considered, what people call the dark night of the soul.

Saint Teresa of Ávila, wrote that outside the palace limits, everything is foul, dark, and infested with toads, vipers, and other venomous creatures. In other words, it is not a graceful or eloquent experience to travel the road to find God. Teresa lived in an age of exploration as well as political, social, and religious upheaval. It was the sixteenth century, a time of turmoil and reform. She was born before the Protestant Reformation and died almost 20 years after the closing of the Council of Trent. Sometimes, when I am jumping over toads and vipers, I catch myself looking out a window. Like today. I am looking for something or someone to help answer the questions of why finding God can be so very difficult. In most cases, I see only my face looking back. Or, if I am looking out from the coffee shop window, like today, and I look just right, I may see a man or a woman staring back at me, and they seem to have the same eyes. I realize we desperately need each other, but sometimes I am unwilling or uninclined or perhaps too scared to reach out.

I am reminded of Dietrich Bonhoeffer. Bonhoeffer was a German pastor, theologian, anti-Nazi dissident, and key founding member of the Confessing Church. He wrote his letters from prison when he was in his darkest of times. In 1944, just before his execution, he explained, "I'm still discovering, right up to this moment, that it is only by living completely in this world that one learns to have faith. . . . I mean living unreservedly in life's duties, problems, successes and failures, experiences and perplexities. In so doing, we throw ourselves completely into the arms of God."

The noise outside the coffee shop heightens the tension. Police are yelling, and passersby are tripping over themselves, trying to take cover. A siren is heard screaming towards us. The noise created by the darkness comes crashing into my soul, making it hard, at times, to breathe. I turn to the ideas of John of the Cross as he eloquently notes that the obscure times of looking for God "crush the silence of our happy little lives." The noise in our ears and hearts drive us from our homes of peace when everything can seem perfect. I would much rather have it quiet and peaceful and beautiful and never sit with the noise of sadness or pain or grief. It is much easier for me to just live and work, and I would rather leave all of the spiritual exercises of finding God to the gurus, the aesthetics, the monks, the nuns, and the white-haired lady in the coffee shop.

There is the sound of the television propelling out ways to make me more comfortable, to make life easier, to drive a better car. There is the noise of my children working on math problems, arguing with their mother, screaming for no reason. Cars outside drive past my home, and if you listen carefully, a passenger jet is humming while it burns its white-hot fuel across the darkening sky. But between these noises, in the cracks, when my four-year-old grabs my hand with her tiny little grip and whispers, "I love you," in these perfectly still moments, I find answers to my fears and I call this God. Rumi understood the importance of finding God in this way but knew the challenge and asked, "Why do you stay in prison when the door is so wide open?"

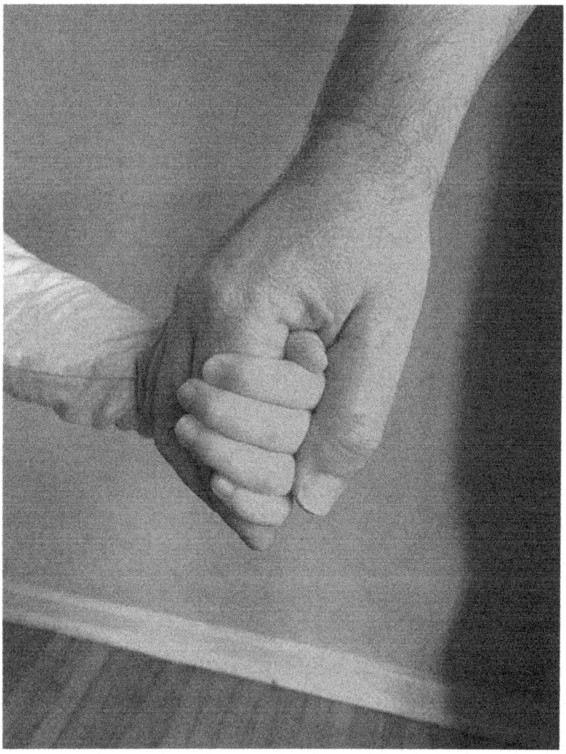

Photo taken by Annette Allan, "Hands," Pacific Grove, CA, March 24, 2020.

Before the old white-haired woman became my traveling partner, it was an ordinary Wednesday in California. The sky was bone grey and the fog raced through the trees as if the trees were reaching to the heavens to comb the hair of God. On this ordinary day, most people had their heads down driving towards the weekend, trying not to look into the cracks, trying to stay clear of the noise.

I sometimes wish I would not always write and think about death. Thinking and writing about death as a Christian theologian is somewhat absurd! Because Christian theology is a theology of life, not death. Or perhaps it could be argued that through death life is discovered. Ray S. Anderson writes that Christian theology is a theology of creation, not corruption; of meaning, not despair; of light, not darkness. Craig Reinhold wrote that the problem of death is the problem associated with being forgotten and it is a human dilemma. Reinhold tells the story of a woman by the name of Bernice Cosulich, a writer and journalist, who committed suicide. In Bernice's suicide letter and in her instructions for after her death, her last wish was to have her work archived and preserved, so, perhaps, "I may not too soon be forgotten." Reinhold argues that her wish is the wish of all humans. "If we die," Reinhold writes, "and all we have spent our lives on is scattered, lost, recycled, unread—what has the toil been for? To be remembered is to preserve our sense that we have done right, ensures that the fragile purpose we have written into our lives survives."

Last Sunday, as I was sitting in church, I was pondering this, and I realized that even Christ suffered from the same fear of being forgotten. As I took the bread and water on that Sabbath day, I recalled Christ's instructions to take the bread and wine to do so in remembrance of his body and blood that was shed for the entire world. Christ knew of his impending sacrifice, his death on the cross. I often feel I should remind people of me. The women on the phone did it, and the man running into the street did it too. I am here, he said, yelling and screaming, *I am here!*

I am still sitting in the middle of chaos, listening to the conversations about life and death. The old white-hair woman just ended her conversation and looks out the window. Then as fast as it came, the fear lifted. It felt like a blanket had been taken off our faces after we were hidden under its cover. Just before I decided to move away from the window and join the lady at her table, to console and to let her know we would be safe, a policeman entered the café with one quick proclamation, "Sorry about that folks, it was a false alarm. Have a great day!" We all shared a nervous laugh. I instantly turned my thoughts away from death, because we were safe. The old white-haired woman with calming hands sipped her hot tea and placed a bite of chocolate muffin into her mouth as if to taste the body and blood of her own existence in this holy place. I stood slowly, folded my laptop lid, placed it in my bag, picked up my drink, slung my bag over my shoulder, and walked to the door. Just before I left, I turned to my fellow travelers and pronounced, "See you all tomorrow."

Lindsay Sledge

My Grandfather's War

On a trip to Texas to visit my parents, I ask my dad if we can open my grandfather's army trunk. I'd wanted to ask for years, but only now on this Thanksgiving morning do I have the nerve. My dad grabs his beer, and we head out to his shop while my mother makes her cornbread stuffing. My dad finds the green, metal trunk hidden under some old blankets and sets it on his work bench. My grandfather, Son, died of a heart attack on my fourteenth birthday. My grandparents didn't have a phone, so my grandmother ran into a convenient store to call 911, while Son died in the front seat of their Suburban. My dad didn't say anything the day Son died. He brought home a birthday cake from Walmart, and we ate it in silence as the blue frosting burned our throats.

It wasn't long after Son's death that the city condemned my grandparent's ramshackle house. My cousin and I spent our summers at their house, pretending we were stranded on deserted islands and fishing for crawdads in the nearby creek. We played unattended for hours until our grandmother yelled at us to come in when it got dark. With their house gone and only a small stretch of grass left in its place, I realize now my imagination accounted for the jungle-like yard of my childhood memories.

My cousin and I begged Son to open the trunk when we were kids, and he'd happily oblige. It sat stowed away in one of the many cars permanently parked in the yard. We'd sit around him, wide-eyed listening to his stories about every item. When my dad finally opens the trunk, it suddenly feels like I'm eight years old again, excited by rummaging through my grandfather's belongings. We begin to sort through the coins from foreign lands, cheap jewelry, and knives. I find Son's whisky bottle that he'd take a swig from only after looking around to make sure my grandmother was still at work. Each item brings Son back. I find an antique stereoscope, and I can almost see him in his suspenders and cowboy hat, showing us the black and white pictures of women with big bosoms inside. He'd let my cousin and I have a quick peek, and we'd say, "Grandpa, where did you get that?" He'd laugh and ask, "This lady's got really big ones, huh?" There was no shame in our looking; we were reveling in being in his inner world. Son collected everything. I find receipts for what seems like anything he ever bought in his entire life, from a car to what I presume was a wedding ring. I find a delayed birth certificate stating his home birth, his death certificate, and everything in between. There is a whole life lived here in this trunk.

I knew Son served in the Army during World War II, but I had no information about what he did during the war. My dad tells me his battalion went to the island of Mindanao to help liberate the Philippines. His Army entrance paperwork says he went to school through the first grade. His occupation is listed as a farmer who planted cotton for his grandfather. Son was illiterate and able only to sign his name. I find many letters in the trunk, and I can't help but wonder if he ever had the pleasure of knowing what they contained. I read one from an Army buddy out loud, in hopes he can somehow hear it.

> "Hello, stranger. Do you remember me or have you forgotten the eight-ball of the first platoon? How's things been going with you? Still punching the cows I guess. How come you never tried riding a caribou in the Philippines? I guess they were a little too wild for you. How'd you finally get out of the Army? I could of gotten out the first part of November, but the Colonel said I was essential so I have to stay a little while extra. I feel like kicking his ass. I have to wait until all the other guys get out to take off. And believe me, I'll take off fast. Until next time, don't drink too much. Your old buddy, Tucker."

Son didn't work when he returned from the war. My mother heard from one of her customers in the beauty shop that Son was the sickest man they'd ever seen alive when he got back, probably from malaria. For years, the talk all over town was my grandfather's unemployment and how my grandmother had to support the family. When the town banker gave her a loan for their Suburban, he didn't hesitate despite their dire financial situation. Instead, he said, "Poor Estelle, always worked to the bone." My grandmother went in early for her shift at the nursing home. She drank her coffee at the table with the other women in the dark hours, to get away from Son. Son was known all over town for being a crotchety old man, but my grandmother got the worst of it. It was never physical, always with words. Son never missed an opportunity to let her know how stupid she was, how awful her cooking tasted, how she talked too much when they were out in public. Son stayed home with their four children, who ran wild all over town. In my dad's elementary school photos, his clothes are neat and pressed, his hair combed. I think this must have been my grandmother's doing. She wanted them to look presentable. While my grandmother worked, Son planted flowers, tinkered in his shed, and collected items from peddlers who stopped by to talk. He'd work an odd job here or there as a plumber, but it was my grandmother's role as provider that held the family together.

Son talked in wild stories, jokes, or insults. My dad doesn't talk much at all, he never has. Instead he grunts here and there or remains silent when you try to talk to him about anything other than music, politics, or food. Here in the shop, as we inspect the items of Son's trunk, I wait and listen for answers. My grandmother has been gone nearly ten years, Son over twenty, and I realize I know very little of what their lives were like. I pull out a wallet to find a small black and white photo of a young Japanese woman. "Who is this?" I ask, knowing it's probably Son's alleged girlfriend my grandmother spoke of. "He met her on R&R in Japan," my dad says. I imagine her still alive, sitting on another continent, wondering whatever happened to her beloved GI. I wonder if the rumored love-child my grandmother spoke of is true. My dad offers no other information and quickly puts her picture back into the wallet. I don't bring her up again. Other pictures of Son in his Army uniform show him young, full of life, a cigarette dangling from his smiling lips. I think he must have been happy to leave his small corner of Texas. Distant great aunts and second cousins tell me Son's father was a harsh, mean old man who once killed Son's beloved hunting dog just for the hell of it. I ask if this is true, and my dad looks off into the distance before handing me my grandparent's wedding picture. They're dressed in simple clothes outside of a rundown house. Their faces are looking straight ahead, the sun making them squint. I'd heard after the war Son lived in a makeshift shelter outside of my grandmother's house until they could get married. I try to imagine them young and in love, ready to start their lives. I cannot picture it. I can only see my grandfather in his own room. He's sitting on the edge of his bed, combing VO5 oil into his hair, listening to Hank Williams on his hi-fi, yelling at my grandmother. She sits alone in her room, which is more of a closet. She cries for the way her life turned out. She cries for the baby they lost and the way he treats her. I can't help but wonder if Son was always so cruel to my grandmother, surely he loved her at some point. I want to ask my dad, but I know his answer is silence. The Christmas after Son died, I bought my grandmother a locket with pictures of him and her. I thought she'd love it, but instead I watched her face as she opened it, and I understood then that she didn't want my grandfather hanging around her neck forever. She was happy to be free of him.

Son never said an unkind thing to me. I was the first grandchild and he loved me. Just as with my father, it was never spoken but it was well understood. On the first of the month when Son cashed his VA check, he'd take my cousin and me to the Snack Shack to get strawberry milkshakes and greasy hamburgers. On the way he'd stop at the utilities office to yell at the water lady for the bill being too high. I tried to apologize with my eyes, but I barely reached above the counter. Son never drove because he couldn't read the driver's test to get his license. All over town my cousin and I walked beside him, him with a cane and us on either side of him. I'd study his wrinkled hands full of age spots and his turquoise ring. Now, here in the shop I look down at my dad's hands as he casually lets the contents of Son's life slip through them. They are the same but different, covered in callous skin and blisters. My dad's hands could build or make anything. They'd never known war. They were the hands of a working man.

As we near the bottom of the trunk, it's the Japanese soldier's flag I want to see now through the lens of an adult. My father finds it and hands it to me. It's stuffed inside a cardboard tube and I slowly pull it out. It's covered in symbols I do not understand, just as I remembered. The symbols I would later find out are names, likely the Japanese soldier's unit, as well as good luck wishes from family. I stretch it out and become mesmerized by its red circle. I try and imagine the scene in which my grandfather came into possession of it. He'd always get quiet when he pulled the flag from the trunk. He told us he killed the soldier and took it off of his body. I didn't understand this killing as a child. Only now do I imagine him young, bent over a dead man, taking this strange souvenir. We'd ask him, "Grandpa, how did you survive the war?" "Jumping over bullets," he'd say. I believed him and pictured him in his full Army gear, running through the jungles, jumping over bullets left and right.

I'm suddenly aware that Son's trunk will be mine one day. What am I to do with this relic from my grandfather's war when it's my turn to take it? How can I honor the horrors of war, my grandfather's suffering, as well as this soldier who my grandfather killed? Would I keep it buried in this trunk, taking it out occasionally to explain war to my grandchildren, as if they won't know what war is. In my research I find an organization that seeks to reunite stolen Japanese flags with their owner's families. I read the story of a Japanese man who was given his dead brother's flag. At a ceremony to return the flag, he wept as they placed it in his hands. Its faint smell brought back his long dead brother. His family never knew the details of his death until the American soldier who'd stolen the flag off of his body provided them. The Japanese family finally found a sense of peace with having a small part of their loved one home. I decide this is what I'm supposed to do with the flag. I feel relief that the burden is not mine to carry into the next generation. I've yet to have this conversation with my dad. I fear he may not agree, and my request will be met with more silence.

As my dad and I go through the last pieces of Son's life, I want to turn to him and shake him and say, tell me what it was like growing up, who took care of you when you were sick, were you happy, did Son ever buy you strawberry milkshakes? I don't say any of this. Instead, we close the trunk and bury it back under the blankets where it came from. I make some small talk and say, "It's getting cold; I better get back in and help with the pies." I wanted answers but I do not find them. Instead, I'm left with more questions. What internal wars was my grandfather suffering from that allowed him to be so cruel to my grandmother? Whatever more my father knows will die here in this shop. I leave the trunk for now, having to make peace with the history I do not know. It will be mine someday, and I will pour over its contents when I am old and think of this day.

Niko Eden

Alarm Red

0400 Day 4

At night the wrestling of air howls. Many times I'm thinking of wolves, the way the air screams through the plastic. Many things are an illusion here, once I climb to bed, it rocks and sways. The gestures of delicate turbulence without the offerings of a cocktail. This is the culmination of the bunk beds we sleep in, a dry country we've deployed into, and some weird apparatus that dispenses air flow in the tents where we sleep. Welcome home.

0430? Day 9

I can feel my shoulder caving its way back through my body like a column of Russian dolls, stacking back into one. The mattress is hard, it is a wonder the troops sleep during quarantine. This is our introduction to Al-Kharj and the weirdness the world has been indoctrinated with in 2020.

Moments before my internal conflict to get up and use the head, I'm disrupted by loud blaring. I'm disorientated, what time is it?

"Alarm Red, Alarm Red."

I look around the tent, bras are out, hair is wild, and their bewilderment is the same.

The tent chief calls out, "Did he say exercise? I didn't hear exercise . . ."

"This is an ALARM RED, I repeat this is an ALARM RED," the big voice announces over the box.

The blaring continues in its appropriate rhythm over the base intercom. We scramble. For the indecent, for some article of clothing to cover. For me, my fucking socks and tennis shoes. My bunkmate and I are struggling at the simplest of functions—God, how long has it been since the first alarm?

We have to start moving quickly now, we wrestle politely out of the tent. Everyone is still in confusion by not hearing the word "exercise." The red dirt kicks up as we find a spot in the bunker. Some were just waking up, some have yet to go to bed and we wait. . .

It is the waiting that haunts us, crammed in a sea of clay bunkers that serve no real protection.

This was not an exercise; a Yemen drone had been intercepted.

Wes Choc

Fore or Aft?

All goodbyes had been surrendered back at port three days ago.

It was 1966.

The hulking ship, a huge beast of greenish gray, lurked silently in the harbor amid randomly exhaled "bwaaaps" and metallic clanks and groans. Floating, going nowhere perhaps, but unanchored, no longer attached to docks by umbilical cords, loading planks, or gangways.

Around where we lingered, close-by reverberations were more muffled inside this new nautical cavity, a cylindrical cavern we were learning to call home. We didn't whisper, but others' voices were still inaudible ten feet away within these humming hollows. Tucked away in narrow, closeted corridors, green metal pipes decorated boxed mechanisms. We speculated about what all these unfamiliar devices did. No one knew, of course, but theories broke the silence. Time passed, pausing only for the next deep-throated "bwaaap" to interrupt.

Some of us lay prone in tight bunks, thinking; some of us leaned against the plumbing or hunched over, retying our shoes. It was private time for us isolated inside our sterile quarters. This was where we waited until the "what nexts" materialized.

Duffle bags had been unpacked and stowed. An hour ago, our stomachs were crammed full of meat and gravy'ed potatoes from endless supplies of slathered pork chops and greasy fried chicken. We were getting used to funny sounds banging above heads or echoing underfoot. Those who had sought onboard places to go had sated their curiosities for the day. Time wasn't particularly precious at the moment. Besides, the more interesting places aboard ship still remaining were off-limits to us, enlisted Marines.

Lolling around San Diego harbor seemed pointless, yet there was no overt eagerness to depart beyond this "let's-just-get-going" attitude that abounded among us, a craving we could do little to satisfy.

Reticence was the norm. But when there *was* conversation, most guys milled around jabbering about food, Tijuana escapades just had, or telling bad jokes . . . y'know, getting-to-know-each-other type small talk . . . the kind that suppresses impatience, using words that surround all the wonderings about what the next "next" would be.

We knew there'd be a John Wayne movie in an hour . . . *Sands of Iwo Jima*. Right outside was our sister ship, the USS *Iwo Jima* . . . so the film was a must-see. But right this minute, bellies full, it was no more than killing time with no place to go.

Again! For the third day in a row. We were getting antsy, suppressing our anxieties.

When it came to understanding war, John Wayne's name often came up, along with his dirty-faced image hugging a rifle, his low voice never saying much . . . all priming us for . . . *some*thing similar! Most of us had no idea of what actually lay ahead, but Marine Corps recruiting posters still summed it all up in our minds . . . minds ready for that *some*thing! So, we demurred along with so many other inevitabilities.

Then it finally happened.

Those powerful, deep, throttled, bass-throated horns announcing departure boomed several repeating "bwaaap . . . bwaaap . . . bwaaap" blasts. It meant the ship was moving, departing San Diego to head west across the Pacific. We all detected less-than-subtle vibrations running up our calves and thighs. Eyes rounded open. There was excitement amid unknowns. Guts tingled. But no Marine would *ever* talk much about such feelings. Our eyebrows lifted and foreheads wrinkled nonetheless.

We dawdled around to gain footing, pretending seasickness or exchanging comments about things we'd do tomorrow. A few guys exited our quarters and clumped up at the bottom of the ladder heading up and out to open-air. Eyes agreeing telepathically, the "let's head up to top deck and see what it looks like outside" intentions were acknowledged with nothing more than nodding heads and pursed lips.

I was one of the last to leave our bunkroom compartment.

In the narrow hall below the ladder up, bunched-up Marines nudged the guy in front with subdued "C'mon!" whispers or "Let's keep it movin'!" nudges. Though slow-going, oddly enough there was no shoving or anger or profanity. As boots hit each ladder rung, there was a muffled clank, a pause, then another clank. Eventually, each guy was given his chance to ascend.

When my turn came, I followed suit, pausing each time as I advanced up one rung at a time through the interior's shaded darkness toward daylight. Looking up near the much brighter opening to the outside world, I could see overcast skies and the legs of a small crowd of Marines. Legs mechanically dissipated to the left or to the right. I finally topped, too. Taking in tropical air, I planted my right foot on the actual deck facing north and hesitated like everyone else had done. My head was being pulled right . . . then left . . . then right again to see *whatever* I could see. My lungs expanded, embracing an already deep breath of humid air.

Dusk was pillowing onto the cloud-laden harbor, but distances were still clear enough in all directions . . . clear enough to capture attentions. Four uncertain Marines lingered nearby; two decisively turned right heading toward the stern of the ship . . . two others toward the bow . . . with head-nods pointing direction but without a single word uttered.

The guy right in front of me frowned left, then right, then left again, squinting toward a low-hanging sun visible below a belly of dark reddish-brown bands and clusters . . . whereupon he shouldered his body right and lumbered back toward tail end to face the flashing lights, reflections of that setting sun on city buildings . . . glass images now miles behind us.

The silent serenity was churchlike. Warm mugginess panned my cheeks as I pushed glasses back up my nose. I straightened my back and adjusted my utility "cover" (Marine jargon for hat). My own pause was peculiar. There were no human sounds, just water swishes from down below. I detected flapping sounds from above, the Stars and Stripes.

In the four or five seconds all this took, the reality of the moment settled into my chest.

The ball of reddened orange relaxed into the skyline to my left, west, squinting as the hazy color-laden sphere hung just below clouds and barely above the horizon. The ship was pointed southwest leaving San Diego, heading straight into the sunset as this new June evening was set to capture one of the longest days of the year.

To the right, a collection of tiny bronzed squares of orange and yellow glinted and flickered off faraway windows. Against a now darkening sky, once formidable skyscrapers were retreating above aprons of distant red tail lights below—an imposing urban American panorama now deserting its dominion but sticking to the retinas of hundreds and hundreds of eyes.

To the left, the clear-cut bloodshot ball kissed the orange horizon silently. The red glow, alluring against blue-green skies and artistically painted clouds, mesmerized its witnesses with a depth of distance far beyond that ocean's edge we could see—a simple yet mysterious horizontal line.

The innards of the USS *Vancouver* groaned subdued guttural engine whines and whirrs while swooshing water skidded by. We gained speed.

Then it struck me. Though we had indeed departed the United States, it wasn't so much about the voyage itself anymore, nor about any warnings or predictions. After all, for months before this June sunset, we had absorbed the destination, endured vigorous training, and digested the never-ending implications of war. Instead, it was indeed about the fore and the aft, this moment at hand, recognition of human choice.

To turn right held comforts for seeing just what was being surrendered, perhaps a warm embrace of those left behind, remembered tears or memories, coupled with acts of stowing these emotions away, maybe for the last time, wishing for an arm around one's own shoulder to identify a wanted inner touch no one would admit to needing.

To turn left tugged the gut, pulling submitted eyes to the beckoning of myriad unknowns—of the "what-might-bes" through uncharted waters in the brain. Yes, and also requiring that same arm around one's shoulder while standing among others so willing.

The pause remained compelling. As had others, I lingered only a couple seconds to glance right over my shoulder, before turning left and pacing toward the bow to ponder what was yet to come *beyond* that summoning red glow.

Leonard Adreon

Some Wounds Don't Bleed

I was sitting in the waiting area of the clinic at the Veterans Hospital in St. Louis when the guy sitting next to me, observing my Korean War veterans cap, started up a friendly conversation. He asked, "Are you really a veteran of that war?" He added that it happened so long ago he had not met anyone who had actually been in the war. I smiled, told him I was 91 years old and that I was with the First Marine Division in Korea in the midst of the war in 1951 and 1952. I was 25 years of age at the time.

He was a good looking young man with curly red hair, freckles sprinkled across his face, expressive green eyes with a continuous tight frown-like expression. I asked him about his veteran experience. He shrugged, told me that he had only recently come home from Afghanistan where the war was dragging on and on. He said he was at the hospital to see if there was someone who could help him cope with his inability to "get my act together and move on with life as a civilian." He described sleepless nights where the slightest noise would cause his heart to race. He wasn't able to lie quietly in the bed. He started drinking and seemed to get some relief through the bottle but knew that was the wrong thing to do. He even tried some Oxy and other stuff but was afraid of what would become of him. He shook his head and said, "I'm a fuckin' mess."

He said, "I'm a lucky SOB. I have had two deployments to that God-awful place and here I am home without a scratch. My best buddy lost his right leg and half his vision from an IED blast. Other friends suffered disfiguring burns and other horrible stuff. You couldn't walk anywhere without worrying that you'd step on a land mine. Why I was spared, I don't know. I feel ungrateful for my good luck." I noticed his eyes cloud up as he choked up getting the words out.

He asked me what my duty was in Korea. I told him that I was a corpsman with the Marines. My job was to do my best to save the lives of those wounded on the hillside battles at the 38th Parallel where we were fighting to take and try to hold the high ground. He wanted to know if I carried a weapon. I told him that I carried an M1 Carbine with a clip that could fire 30 continuous rounds, which was standard issue for corpsmen with the Marines. I also explained that I had a .45 pistol that was recovered from a lieutenant who had died leading my platoon up a hill. He got a kick out of my telling him that because I used to pitch baseball I also had grenades, because I could throw further and more accurately than some of the riflemen.

I found it curious that he asked, "Did you ever kill anyone?" My answer was, "In an assault I fired at the Chinese enemy, but I never knew whether any of my bullets found their mark. I wasn't a very good shot. The Marine fire teams I was with were shooting at the same time. I threw grenades into machine-gun embankments at the same time the Marines were hitting the Chinese machine gunners with bazooka fire and flamethrowers. Even when we succeeded, I never knew whether my grenades did their job."

I hesitated and then continued, "One time, after we had taken a hill, I was ordered to go with six Marines and search some trenches to see if any Chinese were hidden there. Our gunnery sergeant led the way; I followed behind. It was dark. The trench took a sharp turn to the right. I saw flashes of light and heard the staccato sound of rapid-fire burp guns coming from an intersecting trench. I rushed forward to treat two wounded Marines. The rest entered the side trench to go after the Chinese. I saw movement in a trench on the opposite side and crawled in there with my .45 at the ready. My flashlight lit up two Chinese who started to raise their burp guns. I quickly fired seven rounds, heard them scream and fall onto the wet ground. They were dead. The only time I knew that my bullets killed."

He seemed intensely interested in what I had to say. Maybe glad he did not have to share anything more with me about his experiences in his war.

A nurse came out and called his name. He got up, gave me a little wave, and followed her. I smiled at him and said, "Good luck."

After he left, my brief encounter with the young man jarred my memory about several days in Korea. I was stationed at the Battalion Aid Station when my boss, a guy we knew as Doc Koz, asked for volunteers to go to the Easy Med Field Hospital. He explained that we were in a lull and they were swamped at the hospital. Four of us corpsmen took a jeep and drove several miles to the tented hospital, built over what was a rice paddy. I was assigned to a psychiatrist. I didn't know there were psychiatrists in Korea. We were in a large double tent. To my surprise the cots in the tent were full of Marines in there for evaluation. The doctor explained that his job was to find out why the Marine was sent to the hospital and what could be done to deal with the problem. In most cases the Marine performed badly in combat, putting others in jeopardy, or failed to perform the duties of his post. If the doctor decided the Marine just was scared and ran, he would be held over for Captain's Mast, a court-like proceeding to determine if the Marine should be subject to punishment. If the doctor decided the Marine needed treatment for his mental condition, he would probably be air lifted to a hospital ship.

I had no idea how I could be of help. The doctor quietly told me that fighting at the 38th traumatized some. That sometimes it is easier for them to talk to a fellow Marine than to a shrink. "I want to find out what happened to them. Whatever they tell you might help. Suggest you ask them why they think they are in the hospital."

Most were 20 years old or younger. He said, "You are older than they are, that might help."

I told him I'd give it a try. The Doctor brought me to a Marine private sitting in a crude chair next to his bunk. He just sat, not moving.

I told him I was from St. Louis, Missouri, and wanted to make him feel better. He stared at me. I offered to shake his hand. He didn't move. He had a smooth soft face, blond crew-cut hair and sad droopy eyes. I walked to the kitchen and got us cups of red-colored drinks. He took it and drank.

I said I couldn't help him if he wouldn't talk to me. He just stared with no change of expression on his face.

I told the doctor. He just shook his head and took me over to another bunk and introduced me to a rough-looking guy, scraggy-faced, unshaven, dark complexion, maybe Hispanic or Native American.

We started talking. "You a corpsman?" he asked.

I nodded, "Yeah."

"Told the doctor, I'm worn out, wanna go home. Need to go home."

"How long you been here?"

He said, "Six months. Feels like six years. I shot six in a hole. Killed 'em all. They were just kids."

He squirmed, shook his head.

"Can't do it anymore."

He took a deep breath. Tears gathered in the corner of his eyes, "God will punish me. I'm goin' to hell. Ready to end it now."

He described standing on a hilltop, straight up, bullets flying all around, didn't duck, just stood there. "I yelled for them to kill me."

"Damn sergeant grabbed me, pulled me down, I bit him, fought him. He and others threw me on the ground. Someone stuck a needle in me. Now here I am, all fucked up."

I didn't know what to say. I told the doctor I thought he needed a preacher, not a corpsman.

"Adreon", the doctor said, "What happened between you and him was good. He opened up, he somewhat grasps the problem. You helped."

Then the first private called me, "Doc, I wanna tell you something."

I went to where he sat. His eyes filled with tears, his face stretched tight in anguish. He reached up, grabbed his head in his hands. He shook violently.

I sat on the bunk, watching. The shaking stopped. He looked at me. "Left our barricade to go take a crap. Came back. Joe, who had replaced me, had a hole in his neck. The three other guys lay covered in blood. I had told Joe I'd be right back. He told me 'Be careful out there.' Probably his last words."

Tears flowed down his cheeks, "Joe was a good guy, a good guy."

The doctor saw what was happening and headed our way. I said something like, "Not your fault, bullets everywhere, not your fault."

The doctor thanked me and said, "These two need treatment."

I moved away, realizing that not all the wounds of war come from bullets or shrapnel or bombs. Some wounds don't bleed.

Jay Harden

Two Rebels Write Home

[30 Jan 2021]

My wife, Carolyn, and I exchanged letters during the Vietnam War, over 100 in all. Those letters were crucial to my emotional well-being. They stabilized my anxiety and connected me back to my real reason for fighting, especially when she wrote me in a letter of just seventeen words that we were going to have a baby.

Many years later I discovered anew her trove of my letters but didn't read them until I got interested in family genealogy. That curiosity led me to discover the letters of my Confederate great-grandfather, William Harmon Harden. When I found them, I read every one, then decided to read my war letters for the first time. What I found was that he and I wrote to our wives about the same things: family, why we were fighting, and the deeper meaning of what we were doing.

Harmon regularly began his letters this way: "Jane, my dear companion, I seat myself with pen in hand hoping these few lines may reach and find you and our dear little Mollie enjoying the best of health, and also the rest of the family connection. We are all well as common." Then he proceeded to say what was on his mind.

In April of 1864, his troop train headed from Savannah to Atlanta to intercept Sherman's army, stopping to refuel at Milner, just a few miles from their farm. There he saw Jane and their daughter, Mary Zelma (Mollie). Four soldiers left the train and deserted. Despite his fear and longing for home, he later wrote that he chose not to ". . . disgrace myself and you."

Upon arriving in the Vietnam combat zone, I understood what he wrote. I was paralyzed by fear, a fear of confidence and competence, a fear I could not show or discuss with my bomber crew, my peers, or others in the squadron. Everyone else seemed professional, experienced, and fearless.

I wrote to Carolyn and said I was going to the squadron commander and turn in my wings rather than endanger my crew. That, of course, was totally irrational and laughable. We were short of navigators and he would have ordered me to fly anyway. Her words soothed, and her wisdom suggested I take it one mission at a time, one day at a time, even one hour at a time. That I did. And after the first two missions, both over North Vietnam, my superb training kicked in and, over my combat tour of 179 days, I found the tanker and the target every time, on time, and brought my crew safely back to base 63 times.

My great-grandfather and I, two Georgia rebels born and bred, wrote letters a century apart to our beloved wives. Those inked papers were the glue that held two marriages together and survived as records of love in two wars, between two men and women connected by marriage and blood.

Harmon volunteered for the Confederacy in January of 1862 at Griffin, Georgia, leaving behind his wife, Jane, their month-old daughter, Mary Zelma, and his job as a schoolteacher. He followed the example of his father, Hillard Judge Harden, who volunteered to fight in the Creek Indian War of 1836.

Due to short supplies from the army, Harmon wore out his shirts and shoes. So Jane supported him with packages of clothes and food, as Carolyn did for me, that also kept his spirits high.

It is fair to say I volunteered for Advanced Air Force ROTC at the University of Georgia, but it is also fair to disclose that I was encouraged by the believed certainty that I would be immediately drafted if I left school. I had decided to take charge of my future this way and my wife-to-be, Carolyn, agreed. When I graduated in 1965, I awaited orders to active duty, then entered the Air Force as a second lieutenant in June of 1966.

After graduating from navigator flight school in 1967, 1968 was the most eventful year of my life. That year I graduated from bombardier school, then B-52 combat crew training, then moved to New England and got certified combat ready, then left her for a world of war halfway around the globe called Vietnam. I promised I would return to her smile, her laugh, her love, and her arms.

And we promised to write as much as possible. Those were the days before the internet, when letters from home were crucial to my hope. Some letters came with a package, like homemade cookies that arrived stale and crumbled to near powder after a journey of 7,800 miles.

Harmon wrote Jane about sickness. He feared disease more than combat. "Fleas and lice here are as common as flies and gnats in Pike [County]." Later he wrote, "I do not fear the Yankees like I do sickness." He contracted ague (malaria today) and denied he had caught smallpox at "Babe Cardin's." His words were strong: "And I say that whoever told it, contrary to the way I told it, told a pusillanimous falsehood!" Babe Cardin's was probably a place of pleasure near Savannah.

In my case, I was DNIF (duty not including flying) for a few days at the Naval Hospital on Guam due to early indications of what later became a duodenal ulcer. Carolyn wrote me, saying, "I have no news, but I wanted to be close & talk so thus this letter." I wrote her back, describing the tests they gave requiring me to drink a kind of liquid chalk.

Harmon was hospitalized with severe measles. He wanted Jane to visit him at Camp Gordon but warned her about the lice and the vulgarity of other soldiers. He continued to be sick and wrote: "We also hear that Stonewall Jackson got his arm shot off. If that be so, it is bad, but no worse for him than a private."

Jane came to visit him anyway and brought the baby, along with her sister and husband, then accompanied him to the General Hospital in Macon. There he was very lonesome for his combat brothers. He was owed $50 in back pay and regretted volunteering. With nothing to do, Harmon began to study the entire Bible, beginning with Genesis, a prelude to his calling to the ministry after the war. He finally got a furlough in 1863 to go home, but by September, he was back in the Macon Hospital with dropsy (a swelling of the stomach, reflecting heart or kidney problems). He was one of the walking wounded in a hospital with 1,000 patients, so he helped dress the wounds of others.

I well remember Christmas of 1968 on Guam. We had just landed from a combat mission and made it to the last Bob Hope show of that year. We had a reserved place at the foot of the six-foot stage. The sloped grass to my right was full of wounded men on hospital beds and their nurses, a field of glistening white.

I stood up when we sang "Silent Night" and looked directly into the eyes of Ann-Margret. I will never forget that indescribable feeling of personal connection and the tears that leaked beneath my aviator sunglasses. Then I, a Southern Baptist, went to my first Christmas Mass with my copilot, both of us in Bermuda shorts. I wrote Carolyn about it, expressing how that day helped me.

Too sick to sit up, Harmon lay flat and scribbled by candlelight at Fort Thunderbolt outside Savannah, "I have undergone more danger with less fear than I ever thought I could. I have never been in favor of the war. Yet I am in it and I cannot get out honorably now."

Several times Harmon chastised Jane. "I have written three letters before this, and I cannot get an answer. I cannot imagine the reason why you do not write. . . . If I don't get a letter before long I shall shit all over Thunderbolt and then you will have some washing to do that will count."

While I was gone to war, Carolyn wrote me over 100 letters. Due to the mail system, I got none some days, then a prized bundle of pink envelopes smelling of White Shoulders. Mail call was feast or famine. She word-wagged her finger at me when I did not reply as often, reminding me that letters are lifelines of love, regardless of wars or generations.

It is hard to realize how combat changes you at the time, and even harder to express in letters the unconscious accumulation of stress, especially the growing realization that I was personally responsible for delivering death and destruction to people I did not know or hate while I sat bulletproof and comfortable in my ejection seat, 35,000 feet above.

My letters, like his, record personal history. They are lessons in learning. The past unremembered and not understood inevitably repeats itself to the sorrow of descendants. That is the price and the hope of our yesterdays.

Corporal Harden was admitted to Ocmulgee Hospital in Macon suffering from chronic diarrhea. A week later he was in the hospital in Americus where he told Jane of his treatment for deafness from operating artillery at Fort McAllister, Georgia. "The doctor put caustic in my ears to burn them out, and I tell you it liked to make me pee, it hurt so bad." A century later, I found a photo of him with a group of veterans in front of the county courthouse. He is holding an ear trumpet.

In February of 1865, Jane gave birth to their second child and asked him to name her. In his last letter of the war, he wrote her a poem naming their daughter Willie Evalina. Now a sergeant, Harmon was captured in April in Macon, Georgia, became a POW, then was released for home.

I came back to Carolyn in March of 1969. As I exited the plane, I instantly saw her in a bright red maternity dress at the upper right corner of the bleachers. The flash of her smile blinded me with love as I saluted the Wing Commander.

My great-grandfather's letters and mine were our personal weapons against the black odds of war. Because of them, two rebels three generations apart made it home.

Robert Brewer

Save Your Own Life

I had two jobs when an East Brunswick, New Jersey, Navy recruiter seemed bent on making it three. "With your English major degree, you're a cinch for our foreign language school," he said.

At night, I was a photographer with a growing film development and print business I ran from the basement of my parent's house. I had a cool client, George Kemper—and souped George's film to make his prints, freeing him up for more photo shoots.

Afternoons I worked as a janitor in the Rutgers University Neuro Psychiatric Institute. With my friends, Dominic, a WWII vet with a piano tuning business, and Steve, a Viet Nam war vet, we held court in the office of Dr. Droll during our lunch hours, taking turns impersonating the great philosophers. I usually became Sigmund Freud, Dominic imitated Schopenhauer, while Steve impersonated Carl Jung.

One day a man and a woman, announcing themselves as Sarah and Anton, took seats next to me on a long hallway ottoman. Dr. Droll called Anton to his office; Sarah and I exchanged pleasant smiles, and she soon disclosed that she taught theater at Queens College, and since I had a photography business, wouldn't I like to film her students in one of her new productions and if I did, I could drop the prints off at her apartment in Manhattan on East 72nd Street. Or deliver the prints in person at her next meeting with Dr. Droll. Sarah sidled up next to me and put the apartment key in my shirt pocket.

The apartment had a view of the street such that I half expected to see Diane Keaton strolling the sidewalk. It may have been her shimmering and sheer blouse or her thin carmine lipstick, but I withheld from Sarah that I was only in this for the money. I also withheld from her that Anton had written me a letter noting that he was friends with Immanuel Kant, had dated Kant's daughter, and even asked her to marry him; that is, before he met Sarah.

Though I never managed to convey any of this, Sarah returned with another check, which she handed to me at an art gallery belonging to my college friend Greg. She embraced me and whispered bitterly, "You always hurt the one you love."

Relief from this enigma came in the form of a one-way train ticket to the United States Naval Training Center in Great Lakes, Illinois. As the wheels clacked in the post-industrial smog of Gary, Indiana, I realized that Sarah, Anton, Dominic, Dr. Droll, and the tiny East 72nd Street apartment were gone forever.

I exempted Immanuel Kant from this memory sweep, and promised myself if they ever gave us a break in boot camp, I would head to the base library and look him up.

Contributors

Leonard Adreon was in the U. S. Navy from 1944 to 1946 and a corpsman with the First Marine Division in combat in Korea in 1951 and 1952. He is a graduate of Washington University in St. Louis, married, has two daughters and six grandchildren. He facilitates writing courses at the Lifelong Learning Institute at Washington University. In May of 2020 he published the second edition of his memoir, *Hilltop Doc, A Marine Corpsman Fighting Through the Mud and Blood of the Korean War.*

Troy Allan is a father, husband, writer, and a retired U.S. Army Chaplain. He lives and tries to figure out life in Koosharem, Utah with his wife and two children. He is a lover of the outdoors and people!

Jason Arment served in Operation Iraqi Freedom as a Machine Gunner in the USMC. He's earned an MFA in Creative Nonfiction from the Vermont College of Fine Arts. His work has appeared in *The Iowa Review*, *The Rumpus*, *ESPN*, the 2017 *Best American Essays*, and the *New York Times*, among other publications. His memoir about the war in Iraq, *Musalaheen*, stands in stark contrast to other narratives about Iraq in both content and quality. Jason lives and works in Denver. Much of his writing can be found at jasonarment.com

Tiffany Avery has been in the Missouri Army National Guard since 2013 and has deployed once to the Middle East. She earned the rank of Staff Sergeant before her recent graduation and appointment as a Warrant Officer. She is also a recent graduate of Missouri State University's M.A. program in Technical and Professional writing. She currently works as a copy editor at a small publishing company in Springfield, Missouri, and resides in the Kansas City area.

David Blome is a combat veteran of the U.S. Marine Corps who earned a PhD in history and taught at Stanford University before leaving academia for a career in urban education. He began writing fiction last year. So far, his work has appeared in *As You Were* and *The Penman Review.*

Anne Boaden served on active duty with the U.S. Marine Corps from 2006–2015, deploying in support of Operation Enduring Freedom as an AH-1W Cobra pilot in 2011–2012 and 2013. She currently serves in the Marine Corps Reserves and lives in England with her husband and two children. She blogs about her life at: captainleatherneck.com.

Robert Brewer was a foreign language translator and interpreter for the U.S. Naval Security Group. His work has appeared in *The Paris Review*, *Connecticut Review*, *Michigan Quarterly Review*, *Proud To Be: Writing by American Warriors* Volume 9 and elsewhere. He has work forthcoming in *War, Literature and the Arts.*

Randy Brown embedded with his former Iowa Army National Guard unit as a civilian journalist in Afghanistan, May–June 2011. A 20-year veteran with one overseas deployment, he subsequently authored the 2015 poetry collection *Welcome to FOB Haiku: War Poems from Inside the Wire.* He also co-edited the 2019 anthology *Why We Write: Craft Essays on Writing War.* As "Charlie Sherpa," he blogs about poetry at www.fobhaiku.com, and about military-themed writing at www.aimingcircle.org

Retired after working for AAA for over 40 years and a Vietnam Veteran (1965–69), USMC, **Wes Choc** has published two books: *Just Dust*, a personal historical narrative about his very unusual tasks while in Vietnam, and *Inconspicuous*, a biography of a Cold War spy who had two famous parents, a very peculiar upbringing, and who played a consequential role after World War II. He is an active volunteer at a local VA hospital in Tucson, AZ.

Z.S. Diamanti is an author of short and long form fiction. His stories have most recently been published in *Gathering Storm Magazine*, *Fiction on the Web*, and *Literally Stories*. He went to school for far too long and has far too many pieces of paper on his wall. He is a military veteran of OEF and pastor. He and his wife reside in Colorado with their four children where they enjoy hikes, camping, and tabletop games. @zsdiamanti

Niko Eden went from a professional ballroom dancing career, appearing in Assassin's Creed and Sabado Gigante to joining the U.S. Air Force. She received an achievement medal for her tour in Saudi Arabia. Her photography has been exhibited with Santa Monica College, published in *Bahia Magazine*, and IMBD credits for the film *Friend of the World* for still photography.

Milton P. Ehrlich, PhD, is an 89-year-old psychologist and a veteran of the Korean War. He has published poems in *Poetry Review*, *The Antigonish Review*, *London Grip*, *Arc Poetry Magazine*, *Descant Literary Magazine*, *Wisconsin Review*, *Red Wheelbarrow*, and the *New York Times*.

Michael Farwig medically retired from the U.S. Army in 2018 after ten years of service. He served in Afghanistan and Guantanamo Bay. He received the Purple Heart for wounds sustained in Afghanistan. He is now looking to understand that weird experience we call War through writing.

Robert Morgan Fisher is the son of a career Naval Flight Officer. His work has appeared in numerous anthologies, journals, and magazines, including *The Saturday Evening Post*, *Upstreet*, *Pleiades*, *The Arkansas Review*, *Red Wheelbarrow*, *Missouri Review*, *0-Dark-Thirty*, and many other publications. Robert holds an MFA in Creative Writing from Antioch University Los Angeles and is currently on the teaching faculty of Antioch in several capacities. Since 2016, Robert has led the UCLA Wordcommandos, a writing workshop for veterans with PTSD.

Harry Gordon is a former newspaper reporter and teacher. He lives in Long Beach, CA. His work has appeared in the *Concho River Review*, *Southwestern American Literature*, and *Worcester Review*. He was in the Army stationed in Germany from 1964–8.

Jessica M Granger holds a bilingual MFA in Creative Writing from the University of Texas El Paso. She is an Army veteran, serving from 2001–2010 in locations such as Ground Zero post-9/11 and The Balkans. Her work can be found at *Breathe Free Press*, *Fredericksburg Literary and Art Review*, *Molotov Cocktail Magazine*, *SHANTIH Journal*, *Proud to Be* Volumes 8 & 9, and she was recently nominated for a 2021 PushCart Prize for her work in *TheNewVerse.News*.

Jay Harden grew up in the South with a deep sense of freedom, playing barefoot in the alleys and creeks. His mother was a schoolteacher and his father was an agricultural county agent and state legislator. He saw the importance of duty, honor, and public service, flying B-52 combat missions during the Vietnam War. He writes his way to understanding his own experience and sharing what he learned with following generations.

T. M. Hudenburg was a former armor officer 2AD and currently resides and writes near the beach in coastal Delaware.

In a past century **Heikki Huotari** attended a one-room school and spent summers on a forest-fire lookout tower. He's a disabled Vietnam-era veteran and a retired math professor, and has published poems in numerous literary journals, including *Crazyhorse*, *Pleiades*, and the *American Journal of Poetry*, and in three collections. A fourth collection is in press.

Benjamin Inks served three years in the Army and has worked an odd array of jobs, from private investigator to personal trainer. Past publications include *The Little Patuxent Review*, *Ilanot Review*, *Adelaide Magazine*, *Creepypasta*, *Licton Springs Review*, *O-Dark-Thirty Review*, and *Winnow Magazine.*

Billy Jenkins is a fiction and essay writer based in Pittsburgh. A Sergeant Major in the United States Marine Corps Reserve with over 28 years of both active and reserve service, Jenkins deployed to Iraq in 2003 and again in 2006. His essay "Going Home" won the essay award and is published in *Proud to Be: Writing by American Warriors,* Volume 6. His work can also be seen in *Taste Magazine* and his story "Banana Milk and Mystery Meat" was recently featured in *Bricolage's Word Play*. Jenkins is a City of Pittsburgh Firefighter with over 17 years of service. He earned his MFA in Creative Writing from Chatham University in 2015.

Veda Boyd Jones's husband was a veteran of the Vietnam War. Her published works include 47 books and over 600 stories and articles in magazines and reference books.

St. Leger "Monty" Joynes has had 5 previous stories and 3 poems published in *Proud to Be*. His award-winning story "First Day at An Khe" appeared in Volume 1. Joynes is the librettist of a classical music oratorio, *The Awakening of Humanity*. He is the author of 22 books and continues to publish fiction, poetry, and non-fiction from his mountain home in North Carolina. Joynes served in the Army (1964–1966) with the 91st Evacuation Hospital.

Elaine Little is an Army Veteran who deployed to Guantanamo, Bosnia, and Afghanistan. She has published short stories, essays, and articles in magazines and periodicals. She was a first-place winner at the 2019 National Veterans Creative Arts Festival for her essay, "Now It's You," and was a Baker Veteran Scholar at the Longleaf Writer's Conference in 2018. Her TV pilot, *The Relatives* was a finalist for a fellowship at the Stowe Story Lab in 2021.

J.G.P. MacAdam is a disabled combat vet and the first in his family to earn a college degree. His stories have featured in *Passengers Journal*, *Line of Advance*, and *Military Experience & The Arts*, among others. You can find him hunting wildflowers with his wife and son, or otherwise at jgpmacadam.com.

Donna Aycock Meares, native of Atlanta, Georgia, lives with her husband in Grass Valley, CA. With an M.S.W. degree, she did social work, then pursued her writing interests while raising children. Donna's father, Staff Sgt. Carl J. Aycock, US Army, WWII, 29 Oct 1940–30 Dec 1945, served in the Pacific Theater. Her husband, 1st Lt. Arthur Day Meares, US Army, served June 1968–June 1970, with foreign service in 7th Maintenance Battalion, Camp Friendship, Korat, Thailand, where Donna, Army wife, worked with the Staff Judge Advocate as Solatium Payments Assistant, to obtain release of U.S. soldiers imprisoned in Thai jails.

Pat O'Regan has been a life-long writer. His publishing success started in high school with publication of athletic and hunting stories in the local newspaper. Since then, he's published many articles, short stories and poems. The novels and plays have only been self-published. He's had many poems published in small literary journals—*Mississippi Crow*, *Wink*, and *RunMinnesota*.

Kristine Otero's nonfiction publishing credits include, *Passengers Literary Journal*, *Proud To Be* (Volume 8) anthology, *Kleft Jaw Press*, *Still Coming Home*, and *O-Dark-Thirty*. In November 2021 her debut piece of fiction will appear in the anthology *Vice & Virtue*. She served in the Army from 2003–2007, surviving two tours of combat. Wanting to share more than her war, she has begun writing in other genres. Kristine earned her undergraduate degree in psychology. She lives with her dog, Dali.

Carl "Papa" Palmer of Old Mill Road in Ridgeway, Virginia, lives in University Place, Washington. He is retired from the military and Federal Aviation Administration (FAA), enjoying life as Papa to his grand descendants and as a Franciscan Hospice volunteer. Carl is a Pushcart Prize, Best of the Net, and Micro Award nominee.

Joseph S. Pete is an Iraq War veteran with a Combat Infantryman's Badge, an author, and an award-winning journalist. His work has appeared in many literary journals, including *Line of Advance*, *O-Dark-Thirty*, and *Proud to Be*.

Victoria Elizabeth Ruwi is the author of *Eye Whispers*, a book of poetry. She served 9 years in the Army. She earned an MFA in Creative Writing from San Diego State University. Her writing has been published in journals and anthologies all over the states.

Rob Silverman's writing appeared most recently in the *Michigan Quarterly Review* and *O-Dark-Thirty*. He was a finalist for prizes at the *Iowa Review* and *Boulevard*. He served on active duty with the U.S. Army from 1969 to 1972.

Lindsay Sledge is a mother of three, living in New Mexico. Her work has been published in *Kindred Magazine*, *The Ma Books*, *Other Worldly Women Press*, and *Voice of Eve*. She is an Air Force Veteran.

Wes Smith is a fifteen-year Army veteran who served during and after Desert Storm. He holds an MA in English from Indiana University and an MFA in poetry from the University of Massachusetts, Boston. He is the father of five and a grandfather. His work has been published in *Proud to Be* Volume 8, *Veterans Voices*, *The Deadly Writer's Patrol*, and *The Wrath Bearing Tree*. He currently resides in Richmond, Indiana, with his wife and their Dachshund, Camo.

Ruth Snipes grew up in Missouri and joined the military after high school, graduating with a degree in English from the Naval Academy in 2018. She always loved taking photos of her environment as a child (her father was also a photographer), and when she was stationed overseas, she became the ship's photographer. She tries to write and take pictures when she can, even as she is deployed.

J.B. Stevens lives in the Southeastern United States with his wife and daughter. He was nominated for the Pushcart Prize for poetry, was a finalist for the Claymore award, and won Mystery Tribune's inaugural microfiction contest.

Ryan Stovall is a former adventurer, world traveler, and Green Beret. His first poetry collection, *Black Snowflakes Smothering a Torch*, is forthcoming from Woodhall Press in Fall of 2022.

Jonathan Tennis is a graduate of Eckerd College (BA), Norwich University (MSIA), and the University of Tampa (MFA). After serving in the U.S. Army, he moved to Tampa, where he still resides and works as a consultant. His writing has appeared in the *Eckerd Review*, *Military Experience and the Arts*, *O-Dark-Thirty*, *Odet*, *Proud to Be: Writing by American Warriors*, *Sanctuary Literary and Arts Journal*, *Chasing Light—a Burgert Brothers Anthology*, and a festschrift in honor of the poet Peter Meinke.

Aubree Tillett is a librarian, an art historian, and a writer. She earned an MLIS from UW–Milwaukee, an MA in Art History from the University of Arizona, and a BA in Creative Writing and Art History from Valparaiso University. Her husband is a Staff Sergeant in the U.S. Air Force, Security Forces. He completed three deployments and has served ten years, so far.

Christian Tillett is a Staff Sergeant in the U.S. Air Force, Security Forces. He has served for ten years and completed three deployments. His first deployment was to Kuwait with a forward deployment to Afghanistan, his second deployment was to Afghanistan, and his third deployment was to Kuwait. He aspires to complete at least twenty years in the service in order to retire.

Born and raised in North Dakota, **Loretta Tobin** graduated from Minnesota State University–Moorhead with a B.S.Ed. She lives in Everett, Washington. She joined the U.S. Navy Reserve and served as the NMCB 18 admin chief when they mobilized to Al Asad, Iraq. After Iraq, she retired from the Navy Reserve with more than 20 years of service. She recently published two poems in Willowdown Books 2020 anthology, *Poems from the Lockdown*. She has six poems that will be published in the anthology *Our Deepest Calling*.

Peter Trivelas is a Vietnam Veteran who has worked as a television writer/editor and is currently on staff at the David Lynch Foundation where he provides free Transcendental Meditation instruction for veterans. He's working on a first novel and collection of short stories. His work has appeared in *Deadly Writer's Patrol*, *Blood & Bourbon*, *Thema Literary Journal*, *The Sunlight Press*, *Dime Show Review*, and he has two short stories in the *Thank You for Your Service* anthology.

Ben Weakley spent 14 years in the United States Army, including deployments to Iraq and Afghanistan and a tour in the Pentagon. His poetry won first place in the 2019 Heroes' Voices National Veterans Poetry Contest. His poetry appears or is forthcoming in *Ekphrastic Review, Vita Brevis, The Wrath Bearing Tree*, the anthology *Our Best War Stories* by Middle West Press, and the anthology *We Were Not Alone* by Community Building Art Works. He lives and writes in the Tri-Cities of Northeast Tennessee.

As **Ben White** was serving his 22-year military career (in the U.S. Army and the U.S. Coast Guard) and again while he was earning his MFA from the University of Tampa, he thought he was a poet. But he is not a poet at all. Ben is a witness. What he writes is testimony.

Tanya Whitney, retired U.S. Army Master Sergeant, began writing poetry a few years ago as part of her PTSD therapy. Her poetry primarily deals with her military service, but she has also written other pieces. She has individual poems published in several anthologies. She also has published short stories to her credit both nationally and internationally. She is a National Veterans Creative Arts Festival Gold Medal recipient in the Inspirational Poetry Category.

Charity Winters is a graduate of the United States Air Force Academy and Austin Peay's graduate English Literature program. Her work has appeared in various publications including *Proud to Be: Writing by American Warriors, The Red Mud Review, Scintilla, The Report, The Free State Review, The Lutheran Digest, Leatherneck, Proceedings, O-Dark Thirty*, and *The Round Table.*

Valerie Elizabeth Young is a loving mother of two, proud military veteran, talented poet, aspiring writer, and servant leader in her community. She is passionate about her personal growth and development, and loves to express herself through her poetry.

Judges Bios

Bree Pye (Photography) is a queer, disabled veteran and former U.S. Army photojournalist living in Trinidad, Colorado. She holds an MFA from the University of Colorado–Boulder, and her fiction and nonfiction have been published in the *Barely South Review,* Southeast Missouri State University Press' *Proud to Be: Writing by American Warriors,* Volume 7 and Volume 8, *The Journal of Compressed Creative Arts*, *the Center of the American West*, and *Waxing & Waning: A Literary Journal.*

Aisha Sharif (Poetry) is a Cave Canem fellow who earned her MFA at Indiana University, Bloomington, and BA in English at Rhodes College. Her poetry has appeared in *Crab Orchard Review*, *Tidal Basin Review*, *Callaloo*, *Calyx*, *Rattle*, and other literary journals. Her poems have been nominated for Pushcart Prizes in 2019 and 2015. Her book, *To Keep From Undressing*, was published by Spark-Wheel Press in January 2019. She lives in Kansas with her husband and two daughters, and teaches English at Metropolitan Community College in Lee's Summit, Missouri.

Jill Talbot (Essays) is the author of *The Way We Weren't: A Memoir* and *Loaded: Women and Addiction*, and the editor of *Metawritings: Toward a Theory of Nonfiction*. Her writing has been recognized four times in *Best American Essays* and has appeared in journals such as *AGNI*, *Brevity*, *Colorado Review*, *Diagram*, *Gulf Coast*, *Hotel Amerika*, *LitMag*, *Longreads*, *The Paris Review Daily*, and *The Rumpus*. She is Associate Professor of Creative Writing at University of North Texas.

Ursula Villarreal-Moura (Fiction) is the author of *Math for the Self-Crippling* (2022). Her stories, essays, interviews, and reviews have appeared in various publications including *Tin House*, *Catapult*, *Prairie Schooner*, *Story*, *Midnight Breakfast*, *Gulf Coast*, and *Bennington Review*. Her writing has been nominated for Best of the Net, the Pushcart Prize, Best Small Fictions, and one of her stories was listed as distinguished in *Best American Short Stories 2015*.